PROCESS SAFETY

An Introduction

The Author

Bob Skelton spent thirty years in the process plant contracting industry. As Manager of Process and Safety for Davy Nuclear he worked on the development of safety documentation for major projects at BNFL Sellafield and for the Sizewell Nuclear Power Station.

Bob is currently the Nuclear Energy Teaching Fellow in the Departments of Engineering and Chemical Engineering in the University of Cambridge. He lectures in safety to undergraduates and organizes the teaching of chemical engineering design. He is Safety Officer for the Department of Chemical Engineering and is a member of the University Committee for Safety.

Bob Skelton also participates in the IChemE Continuing Professional Development programme. He is a member of the IChemE Register of Safety and Loss Prevention Specialists.

RELATED BOOKS PUBLISHED BY IChemE

Hazard identification and risk assessment
Geoff Wells
1996, £68.00, ISBN 0 85295 353 4

Risk assessment in the process industries
Robin Pitblado and Robin Turney (eds)
1996, £28.00, ISBN 0 85295 323 2

Developing effective safety systems
Ian G Wallace
1995, £33.00, ISBN 0 85295 358 5

Lessons from disaster: how organisations have no memory and accidents recur
Trevor Kletz
1993, £28.00, ISBN 0 85295 307 0

Hazop and Hazan
Trevor Kletz
1992, £27.50, ISBN 0 85295 285 6

An engineer's view of human error
Trevor Kletz
1991, £26.50, ISBN 0 85295 265 1

Process plant design and operation
Doug Scott and Frank Crawley
1992, £22.50, ISBN 0 85295 278 3

Human factors in process operations
Robert C Mill (ed)
1992, £22.50, ISBN 0 85295 294 5

PROCESS SAFETY ANALYSIS

An Introduction

Bob Skelton

INSTITUTION OF CHEMICAL ENGINEERS

Published by
Institution of Chemical Engineers,
Davis Building,
165–189 Railway Terrace,
Rugby, Warwickshire CV21 3HQ, UK.
IChemE is a Registered Charity

© 1997 Bob Skelton

ISBN 0 85295 378 X

Printed and bound in Great Britain by CPI Antony Rowe, Chippenham and Eastbourne

To Patricia and Fiona

PREFACE

Safety has always been paramount in the minds of those responsible for the design and operation of process plant. Until the 1960s most organizations had their own methods of safety assurance, relying very much on in-house experience and expertise. The whole structure of the process industries has changed dramatically in the last thirty years and much of the accumulated experience and expertise has now been lost. It is therefore essential that formal methods of safety analysis are available to the process industries to ensure the safe and efficient operation of their processes. The changes in legislation — driven initially in the UK by the Health and Safety at Work etc Act 1974 (H&SAWA) and more recently by the European Union — and with the general trend towards a more litigious society, have encouraged the use of more formal safety assessment methods that satisfy the regulators and can, if necessary, be defended in court. In the USA the Occupational Safety and Health Act 1970 (OSHA) places similar obligations on employers as the UK H&SAWA.

The literature on formal safety assessment methods has grown considerably. There are now several professional journals devoted to the topic, a number of advanced general text books and a large number of books on specific subjects. The range of books written or co-authored by Trevor Kletz, including *Hazop and Hazan*, *An Engineer's View of Human Error* and *Lessons from Disaster* (see inside back cover), form a particularly valuable contribution to the literature. When I moved from industry to the academic world, however, I could not find a suitable, affordable, undergraduate textbook which covered the requirements of the Institution of Chemical Engineers (IChemE) accredited safety syllabus. I therefore started to produce copious lecture handouts to fill the gap. Now this book is intended to do the job.

Though intended as an undergraduate textbook, the book should also be of value to all engineers involved in safety, and particularly to those who have not had any formal training in the techniques of safety analysis. The basic techniques described are also of interest to members of the public, local authorities and others who are concerned with the safety of the process and related industries. To that end some material has been included which is not strictly part of a typical undergraduate course.

There is always a debate about how safety should be taught in an undergraduate course, one of the problems being to teach in such a way that it is readily examinable. The result is sometimes to concentrate on the more mathematical aspects, neglecting those techniques which involve some degree of judgement. This is understandable given the lack of experience of many undergraduates, but it can leave them with an incomplete knowledge of the techniques involved. This book aims to show undergraduates (and others) how safety assurance is actually performed in industry.

With this in mind the book moves from hazard identification to frequency and probability analysis and only then does it consider consequence analysis. The result is that mathematical treatment is not introduced until halfway through the book. I make no apology for spending a considerable amount of time on the technique of Hazop; it is by far the most important tool used in process plant safety analysis and in many cases is the only one needed. The inclusion of Failure Mode and Effect Analysis (FMEA) in a book aimed primarily at chemical engineers may be questioned but I feel that it is also a useful tool for the process industries.

Chapter 3 on safety in design and operation has been placed where it is in order to put all the techniques used in perspective and to stress at an early stage the fact that safety must always be proactive, not reactive. The section on human factors goes much further than is needed in most undergraduate courses, but it is an area of increasing importance in safety assurance and thus merits a comprehensive treatment.

The mathematical treatment has deliberately been kept simple. It is not my intention to give rigorous derivations of all the formulae used in safety analysis. Readers requiring a fuller proof of the methods discussed are advised to consult more detailed works, such as those by Frank Lees.

One of the main problems in any form of quantitative safety analysis is that of obtaining the required data. After all, the analysis is only as good as the data employed. I am grateful to Mechanical Engineering Publications for giving me permission to reproduce information taken from its publication *Reliability of Mechanical Systems*, 2nd edition, 1994, and I would suggest that this work is taken as a starting point when seeking data. Because of the problem of ensuring applicability of data, readers are always advised to go back to original references rather than copy data straight from this or any other textbook.

Because this book is essentially a textbook for undergraduates, I have included example problems (see Chapter 11). Some are taken from the University of Cambridge Chemical Engineering Tripos (with the permission of the University) and others have come from the wide range of safety courses run by Jenbul Ltd for the IChemE. The former are designed to be completed within 30 to 40 minutes, whereas some of the latter are open-ended and may take longer.

My background in safety has been largely as a contractor to the nuclear industry, where many of the techniques (except Hazop) originated. The nomenclature and presentation may therefore, at times, be slightly biased towards that industry. The methods described are, however, now widely used throughout the process industries.

Bob Skelton

ACKNOWLEDGEMENTS

I would like to place on record my thanks to the many people and organizations who have encouraged and assisted me in this venture. It is difficult to list everyone but particular mention must be made of my colleagues in the Chemical Engineering Department and Magdalene College at the University of Cambridge for their encouragement, Colin Bullock of Jenbul Ltd for his many suggestions, his assistance in proofreading and for consent to use Jenbul material, my former colleagues at Davy International where I learned and practised the art of safety assurance for many years and several members of the Health and Safety Executive for their helpful comments on the overall content and on certain sections, particularly Chapter 6 on risk.

Finally I must thank my wife Patricia for her forbearance during the many hours which writing this book has consumed and for her assistance with some of the proofreading.

Bob Skelton

CONTENTS

1. INTRODUCTION

1.1 THE IMPORTANCE OF SAFETY

All engineers have a duty to use their best endeavours to ensure that plant designed and/or operated under their control is as safe as is reasonably practicable. Not only is this now a legal duty[1], it should also be considered as a moral obligation.

This duty is wide reaching and can be divided into a number of topics as follows:

- prevention of death or injury to workers;
- prevention of death or injury to the general public;
- prevention of damage to plant;
- prevention of damage to third party property;
- prevention of damage to the environment.

Whilst prevention of death or injury to human life must take top priority, do not take this list as a general indication of the order of priority. Many people today would put protection of the environment above that of property. The term 'loss prevention' is sometimes used to cover the five topics and also the general economic loss, such as loss of market share and loss of goodwill, that can result from accidents and other untoward incidents. The current trend is to link safety, health and protection of the environment and treat them as an integrated topic[2].

In most industries, the main concern is to ensure worker safety by such things as machine guards, moving load warnings and electrical isolation. Accidents (fire excepted) rarely have any effect on other workers or members of the public.

The process industries are in a different situation, however, because accidents can result in the release of toxic materials or large amounts of energy with disastrous consequences for workers and third parties. Releases from a chemical plant can go well beyond the site boundary and can cause both long-term and short-term effects. The problems resulting from incidents such as Flixborough, Seveso and Chernobyl impose a societal as well as an individual dimension. Note, however, that even in process industries handling very dangerous materials, the majority of accidents are not related to processes — they are largely trips, falls and dropped loads.

Much can be done to ensure safety by the simple application of common sense and basic engineering skills. As processes become more hazardous, however, the problems of ensuring safe operation become more complex, requiring the application of specialist safety analysis methods. Such techniques can only be acquired by specific training and experience.

1.2   TERMINOLOGY

It is important at the outset that there is a clear understanding of the various terms used in safety work. The following definitions have been developed by the IChemE[3].

Hazard is a physical situation with a potential for human injury, damage to property, damage to the environment or some combination of these.

Risk is the likelihood of a specified undesired event occurring within a specified period or in specified circumstances. It may be either a *frequency* (the number of specified events occurring in unit time) or a *probability* (the probability of a specified event following a prior event), depending on the circumstances. It is often expressed in mathematical terms involving both failure and consequence. Thus if a vessel containing a toxic chemical ruptures, it will only pose a hazard if people or property are in range of its harmful effects. If there can never be any people within range then the consequence and thus the risk to people is zero. There may, however, still be a risk to property or the environment. It is frequently split into *individual* and *societal* risk. *Individual risk* is the frequency at which an individual may be expected to sustain a given level of harm from the realization of specified hazards. *Societal risk* is the relationship between frequency and the number of people suffering from a specified level of harm in a given population from the realization of specified hazards.

Hazard analysis is the identification of undesired events that lead to the materialization of a *hazard*, the analysis of the mechanisms by which these undesired events could occur and usually the estimation of the extent, magnitude and likelihood of any harmful effects.

Quantitative (or Quantified) Risk Assessment (QRA) is the quantitative evaluation of the likelihood of undesired events and the likelihood of harm or damage being caused, together with value judgements made concerning the significance of the result.

1.3   SAFETY ASSURANCE

The complete elimination of risk in human activity will never be possible; all that can be done is to ensure that the risk is kept as low as reasonably practicable. It is stated in Section 1.2 that risk is a function of both the frequency or probability of an untoward event and the consequences of that event. Thus it is possible to consider safety assurance from both aspects. It is generally agreed that the best way is to reduce the consequences, and this can best be done by reducing the potential for harm of particular operations. To quote Trevor Kletz, one of the best known safety professionals in the UK, 'what you haven't got can't leak'[4]. This concept is often termed 'inherent safety'. It must be accepted, however, that in the process industries inherent safety will never be totally possible, so other methods of safety assurance are required. Such methods are the subject of the greater part of this book.

Though not written specifically for the process industries, the UK Management of the Health and Safety at Work etc Act Regulations 1992[5] summarize the concept of safe design and operation very well as follows:

• if possible, avoid the risk altogether — for example, by not using a particular substance or process;

• combat risks at source rather than by palliative measures;

• adapt work to individuals — for example, apply good ergonomics;

• take advantage of technological progress;

• include risk prevention as part of a coherent policy;

• give priority to measures which protect the whole workplace;

• ensure everyone understands what they need to do;

• ensure the existence of an active health and safety culture throughout the organization.

Of these points, the first and the last are probably the most important.

The importance of safety assurance is also clear in Section 5 of the US OSHA[6] which states that each employer shall furnish to each of his employees employment and a place of employment which are free from recognized hazards that are causing or are likely to cause death or serious physical harm to his employees.

1.4   SAFETY CULTURE

A company will only have a good safety record if it has the right attitude to safety — that is, a good safety culture. This must start right at the top of the organization and continue down to the lowest grade of worker on the shop floor or site. The attitude of top management is particularly important as it can, and should, have an influence on everything that goes on in the organization. It is also vitally important that this attitude is correctly perceived by the workforce. Studies in a number of companies have shown that if the shop-floor workers think that the top management cares about safety then the whole organization will have a good safety record. Comparisons of different plants within the same industry have also shown a clear correlation between plant safety and the perceived attitude to safety of the local management.

The attitude of top management to safety must be continuous and interactive. It is not possible to achieve a good safety culture simply by appointing a safety officer and leaving everything to him or her. Safety must be as important an item on board meeting agendas as economic matters. Without top level backing, it is difficult to ensure that provision is made in company procedures and project programmes for the necessary safety analysis. It is also difficult for safety staff to get any changes made when under pressure from project or production staff unless they know that they can rely on support at the highest levels.

Historical factors play an important part in the development of a good safety culture. In general, the newer industries and those handling the most dangerous substances have the best safety cultures. Thus the nuclear industry has probably the best safety culture in the country and the larger chemical companies also have excellent records. One of the most dramatic improvements in safety culture was in the coal industry after nationalization. Partly because the industry had historically been considered dirty and dangerous, it used to have a poor safety culture and a bad accident record. Thanks to new technology which changed the image of the industry, and also to a committed top management, the industry changed its safety culture and this resulted in a dramatic fall in the number of accidents. Another example of a change in safety culture is the UK offshore oil and gas industry. Following the Piper Alpha accident much has been done to improve the attitude to safety at all levels within the industry.

1.5   SAFETY ASSESSMENT

There are a number of techniques available to engineers and safety specialists to ensure that all has been done to assess the level of risks associated with a particular operation. In order to do this the following questions must be asked:

- what can happen?
- how often can it happen?
- what are the consequences?
- is the risk acceptable?

The first question is answered by hazard identification techniques such as the hazard and operability study (Hazop) and Failure Mode and Effect Analysis

(FMEA). The second and third questions are answered by a variety of Quantitative Risk Assessment (QRA) methods such as Fault Tree Analysis, Event Tree Analysis, dispersion modelling and so on. The final question is much more complex, involving social and political considerations as well as technical matters.

1.6 CONCLUSIONS

Whilst nothing can guarantee absolute safety, provided that everyone in an organization realizes that they have a moral and often a legal duty to ensure the safety of the operations in which they are involved, then it should be possible to ensure that the risks are kept as low as reasonably practicable.

REFERENCES IN CHAPTER 1
1. Health & Safety at Work etc Act 1974 (HMSO, London, UK).
2. Turney, R.D., 1990, Designing plants for 1990 and beyond: Procedures for the control of safety, health and environmental hazards in the design of chemical plant, *Trans IChemE*, 68 (B1): 12–16.
3. Jones, D.A. (ed), 1992, *Nomenclature for Hazard and Risk Assessment in the Process Industries*, 2nd edition (IChemE, Rugby, UK).
4. Kletz, T., 1984, *Cheaper, Safer Plants or Wealth and Safety at Work — Notes on Inherently Safer and Simpler Plants* (IChemE, Rugby, UK).
5. Management of the Health and Safety at Work etc Act Regulations 1992, *Approved Code of Practice* (HMSO, UK).
6. Occupational Health and Safety Act 1970 (OSHA) (as amended 5 November 1990) (US Dept of Labor, Washington, DC).

2.   THE CONCEPT OF RISK

There is nothing new in the concept of risk management. From the start of time all creatures have had to run risks to survive. In the animal kingdom it is often a matter of taking the risk of being eaten by something else or not eating at all. *Homo sapiens* has developed beyond this stage of existence but still must use intelligence to make judgements in order to survive the very different risks which are faced in today's world. Man has learned to recognize risks, to accept them by weighing the various risks involved and at times to take risks even to the point of gambling. Thus throughout the history of man there has been a process of education and training in risk management.

2.1   DEFINITIONS OF RISK

Chapter 1 gives one formal definition of risk. Whilst this definition may be accepted by safety practitioners, and to some extent by the courts, there is still a great deal of confusion as to what is meant by risk. A recent publication by The Royal Society[1] quotes ten formal definitions of risk or riskiness, including the one already given (page 2). All ten definitions imply some degree of quantification, though it must be stressed that they involve a considerable degree of judgement. It is thus understandable that attempts to discuss the quantification of risks with the general public are fraught with difficulties.

The IChemE's definition — that is, a function of probability (or frequency) and consequence — is sometimes termed 'expected loss'. A slightly different form, 'the probability of undesired consequences', has been adopted by The Royal Society[1]. For the purposes of this work, the IChemE's definition is generally used, as it leaves the way open for a two-pronged attack on risk reduction. Thus later chapters look at ways of assessing both probability (or frequency) and consequence. Whatever definition is used, it must be realized that a hazard does not automatically result in a risk.

Frequency is normally expressed in terms of 'events per year' and probability is a dimensionless number between 0 and 1. There are a number of ways of expressing consequence; the two most common are simple monetary sums and the number of fatalities. The latter is used in most risk comparison statistics because it is so absolute. Sometimes the term 'equivalent fatalities' is used but this can be difficult because, as will be discussed later, some accidents can result in 'fates worse than death'.

Risk can occur in a number of different ways, not all of them obvious to those exposed. Risk can thus be further broken down as follows:

Unknown risks — risks of which the consequences are either not known or not fully understood, a good example being the risks perceived to be associated with nuclear power and with genetic engineering techniques.

Concealed or unconscious risks — risks which are not obvious or the hazards are not readily detected by the senses; situations in which people may be lulled into a false sense of security. Ionizing radiation is an example of a hazard which is not detected by the senses.

Conscious risks — risks of which people are aware; they can be detected by the senses giving the exposed person some warning and perhaps the ability to take mitigating action. Examples are the risk of getting burned in the event of a fire or of falling off a rock-face when climbing.

Predictable risks — risks of which both the probability and the consequence can be estimated with some degree of confidence. This can allow some form of judgement to be made by those exposed and those responsible for their protection.

Temporary risks — risks which are present only for a short period of time and may be acceptable only for that reason; an industrial example is the risk involved in some maintenance operations, a more mundane example is crossing a busy road.

Calculated risks — predictable risks which are taken knowingly because there is a large perceived gain, because of special circumstances such as rescue or maintenance or because the risk is part of the attraction, as in dangerous sports.

Each type of risk demands a different approach, both in analysis and in setting criteria for acceptability.

Another way of classifying risk is to look at the probability/consequence relationship. Thus risks can range from high probability/low consequence to high consequence/low probability. The latter tend to be more difficult to assess and comprehend.

2.2 ACCEPTED AND IMPOSED RISK

Before considering what is acceptable, a distinction must be made between risk that is willingly accepted and imposed risk. Workers in dangerous industries are well aware of the risks involved and receive special training and often medical surveillance. They are prepared to accept the risks, usually for enhanced rewards. Similarly, participants in dangerous sporting activities accept the risks involved as part of the price to pay for the excitement or money they obtain from that activity. People in both categories are prepared to accept a much higher degree of risk than members of the general public who may be subjected to danger from the same activities. Thus three categories of risk are normally considered when examining industrial activities:

- occupational risk to the workforce;
- individual risk to the general public;
- risk to society.

OCCUPATIONAL RISK TO THE WORKFORCE

This category is self-explanatory and can usually be assigned a higher risk value than the other categories.

INDIVIDUAL RISK TO THE GENERAL PUBLIC

This is the risk, usually of an immediate consequence, to individual members of the general public from an untoward event. Such a risk is both person- and location-specific and can be defined as:

Risk = (frequency of event) × (casualty probability) × (fractional exposure)

The casualty probability is the probability of the untoward event causing a fatality and the fractional exposure is the fraction of time for which a person is likely to be present at the location in question.

Where there are several untoward events or potentially hazardous locations, the overall risk is the summation of the individual events and location risks.

SOCIETAL RISK

The concept of societal risk reflects the likelihood of accidents involving multiple casualties and/or long-term detriment including, for example, contamination of the environment. Thus the effects on society from accidents such as Chernobyl and Seveso persist long beyond the immediate effects on those individuals directly involved.

This leads to the need to examine the relationship between individual and societal risk. Any risk within an area surrounding a dangerous process depends upon its frequency and consequence not only to the individual but also to the number of individuals likely to be involved.

The consequences of bad siting of process plants were shown dramatically at Bhopal and following the LPG explosion in Mexico City. A very thorough study of the consequences of siting process plant in heavily populated areas in the UK was conducted by the Health and Safety Executive (HSE) some years ago on Canvey Island[2]. As a result, changes were made on the site to reduce the level of risk to the local population.

2.3   PERCEPTION OF RISK

The question of acceptability of risk is complicated by the problems of perception. To understand this topic fully it is necessary to have some knowledge of psychology, sociology and anthropology. But because it is often left to the engineer or safety specialist to win acceptance for a new project, it is essential that they have some idea of the concept of risk perception. This is particularly important because of the differences which frequently exist between the perceptions of experts and laymen[3].

Many surveys have been carried out over the last twenty years to determine the various factors which affect the perception of risk, and a number of models have been proposed[1] to explain the findings. Because of the many psychological, anthropological and sociological influences involved, it is not really surprising that no one clear model has been found. The various studies do, however, have a number of factors in common. It has been proved beyond all reasonable doubt that the following factors result in a perceived increase of a risk:

- involuntary exposure;
- lack of personal control over outcome;
- uncertainty of outcome;
- lack of personal experience of risk;
- delayed effects;
- genetic effects;
- low frequency/high consequence events;
- human rather than natural causes.

Consideration of these factors can explain several well-known faulty perceptions by laymen and the sometimes irrational demands for very high and inconsistent safety standards.

For example, the public throughout the Western world demands standards of safety for rail travel several orders of magnitude higher than those for travel by road. Most car drivers, often mistakenly, think that they have personal control whereas rail passengers are in the hands of others. All car drivers have at some time or other been involved in an accident or witnessed one, whereas very few rail passengers have ever been involved in or witnessed a rail accident. Road accidents happen with great frequency but the consequences are usually low; it is rare for more than two people to be killed in a road accident. Rail accidents are very rare but the occasional accident sometimes results in a significant death toll.

A similar comparison can be made between nuclear power generation and hydro-power. Statistically hydro-power is much more dangerous but the public perception is the reverse. The problems in this case relate to uncertainty, delayed effects, perceived but not proven genetic effects and the fact that dam failures — although usually due to bad design or construction — are seen to be due to natural causes because the initiating event is usually natural.

A further factor influencing both rail travel and nuclear power is that the public generally expect much higher standards from public or other large corporations than from small private companies.

Another approach[4] to risk perception uses three factors:
- dread risk;
- unknown risk;
- number of people exposed.

This is often called the psychometric approach and the results are presented diagrammatically. The 'dread' factor includes such things as personal control, risk to future generations and whether or not it affects the person concerned. Nuclear power scores particularly poorly on this method whereas domestic appliances — which are the cause of a large number of accidents each year — are seen as benign.

There is also an aversion factor which must be considered; people have a particular and understandable aversion to the prospect of dying from cancer related illnesses. Hence they have a greater aversion to activities which could possibly result in cancer than to those which may bring about sudden death by, say, the bursting of a dam.

The matter is further complicated by social, cultural and personality factors. Some people are born gamblers and risk-takers whilst others are naturally very cautious. Some are individualists with little care for anyone else or the future whereas others have a greater social awareness. Another group of people can be classed as fatalists who accept everything as fate or the will of God. Attitudes to risk vary considerably.

Some people are prepared to listen to and trust the experts and regulators whilst others have no faith in them whatsoever. It is very important that people who are in a position to influence public opinion on matters of risk and safety are careful not to say anything which can reduce the credibility of the profession. Unfortunately, the general mistrust of politicians by the public is not making the life of the regulator any easier.

The media can also play a large part in affecting the way in which the general public perceives risk. The fact that a minor rail accident or a totally non-nuclear incident at a nuclear power plant gets wide media coverage, whereas a road accident generally has to kill more than four people before reaching the national media, shows how the media perceives risk. The media has, however, done much to condition public thinking on matters of risk and public safety, much of it positive.

Attitudes of the general public are, however, complex, partly because of the lack of knowledge and partly because of fear and prejudice. Most members of the public find risk analysis a very difficult area to understand and are difficult to convince by logical mathematical argument.

Finally, reverting to the process side of the fence, it is important to remember that the operators also tend to exercise differing attitudes to perceived risks. The exercise of overseeing process safety must always accommodate an awareness that each risk scenario tends to be seen and judged differently by those who are exposed. This matter is considered in more detail in Chapter 9.

2.4   QUANTIFICATION OF RISK

Though it has been shown that, in the eyes of the layman, risk is largely subjective, none the less it is important to use some form of quantification — if only for comparison purposes. A considerable amount of statistical data on risks arising from common activities now exists and it must be expressed in a form that is useful, not only for safety professionals, but also for the general public and for decision-makers.

INDIVIDUAL RISK DATA

For individual risk the simplest way is to express the data as the risk of death in any one year due to a particular activity. Even this simple approach is complicated by the question of exposure time. Table 2.1 on page 12 gives some mortality statistics for the UK on this basis.

Whilst this data can be used for risk comparison purposes, it must be used with caution. It is now generally accepted[5] that it is wrong to compare the

TABLE 2.1
Level of risk per year

Smoking 30 cigarettes per day	1 in 200
Man aged 35–44	1 in 600
Motor vehicle accident	1 in 10,000
Accident at home	1 in 12,000
Accident at work	1 in 30,000
Rail accident	1 in 420,000
Terrorist bomb (London)	1 in 5,000,000
Lightning	1 in 10,000,000
Animal venom (mostly wasps)	1 in 20,000,000

risk of death from a wasp sting to that due to the release of ionizing radiation from a nuclear power plant accident, even if the numerical value is the same.

A generally more useful statistic for comparison is based on death per unit of activity. This takes into consideration the exposure time to the hazardous activity. The most frequently used statistic of this type is the fatal accident rate (FAR) which was originally developed in the UK chemical industry but has now been extended to cover a wide range of industrial and other activities.

The FAR is the number of deaths expected in a workforce of 1000 people during a working lifetime (10^5 hours). It is thus the expected death rate per 10^8 exposed hours. Table 2.2 gives some FAR data for industrial occupations and Table 2.3 gives data[6] for other activities.

Comparison of the two tables shows the most dangerous part of North Sea oil and gas operations is the helicopter travel to and from the rigs, even though the statistics include the Piper Alpha tragedy. It shows that even an able-bodied man is as likely to be killed in an accident at home as at work because on average he will spend only 20% of his life at work. The two tables also show very clearly the risks people are prepared to take in leisure activities compared with other activities.

The question of exposure time is still not totally resolved, particularly for leisure activities. The figure for rock climbing is based on time actually spent on the rock-face and hang-gliding is time spent taking part. Sporting figures are often given on the basis of 10^6 participant hours.

The best way to compare records for travel accidents is to use the distance travelled, usually 10^9 km, as shown in Table 2.4. This method favours air

TABLE 2.2
FARs for industry

Chemical industry	2
UK industry (factory work)	4
Coal mining	8
Deep sea fishing	40
Offshore oil and gas	62
Steel erectors	70

TABLE 2.3
FARs for other activities[6]

Terrorist bomb in London area	0.01
Living close to a nuclear plant*	0.1
Staying at home (able-bodied men)	1
Staying at home (all groups)	4
Rail travel	5
Car travel	30
Air travel (UK internal)	40
Smoking (average)	40
Pedal cycling	96
Helicopter travel	500
Motor cycling	660
Rock climbing	4000

* This is considered to be a tolerable limit based on a 1 in 100,000 event

TABLE 2.4
Risk per 10^9 km travelled (UK only, 1986–90)[6]

UK scheduled airline passengers*	0.2
Rail passengers†	1.1
Car drivers and passengers	4.4
Pedal cyclists	50.0
Pedestrians‡	70.0
Motor cycle riders	104.0

* Excludes Lockerbie (terrorist incident, non-UK airline)
† Includes King's Cross underground station fire
‡ Assumes 8.7 km per person per week

travel because most accidents occur during take-off or landing. The rail figure is increased by the King's Cross underground station fire which was not strictly a transport accident. The air travel figure is for UK airlines only, so the Lockerbie accident is not included. Air travel statistics normally exclude terrorist incidents.

The period over which data are taken can be very important. Many statistics use a five-year rolling average so one accident can have a significant effect. This is particularly so in low frequency/high consequence situations such as air and rail travel but less so in other areas. The equivalent UK rail accident figures for other five-year periods average 0.55 — that is, half the figure in Table 2.4 — whereas the road accident figure was about the same.

Industrial accident statistics have generally shown a significant improvement over the last forty years. The coal and chemical industries show particular improvement although some, including the construction industry, have shown little or no improvement. Road accident figures have fallen recently due to seat belt legislation, stronger drink-driving law enforcement and better vehicle design.

SOCIETAL RISK DATA

One way of expressing the risk to society is to consider it in terms of the F/N curve in which the frequency of events F causing N or more deaths is plotted against the number of deaths N.

Figure 2.1, taken from Reference 1, shows a mixture of actual and estimated F/N curves for a wide variety of activities and is based on both US and European data.

Whilst mortality statistics are the simplest to compile and compare, they do not give the total picture. The 'fate worse than death' concept looms large in the minds of the general public and tends to be reflected in court settlements. Thus permanent brain damage or total paralysis is generally considered to be worse than death and this fact should be allowed for in comparisons of risk. It is, however, very difficult to quantify 'fates worse than death' except on the basis of court awards, which can vary considerably for what seem to be very similar injuries. The questions of economic loss and cost benefit are discussed in Section 2.7, page 20.

2.5   ACCEPTANCE CRITERIA

The decision about what degree of risk is acceptable must, in the end, be made on political grounds. It is not the job of engineers or safety practitioners to make this sort of decision but they must provide the information so that a rational

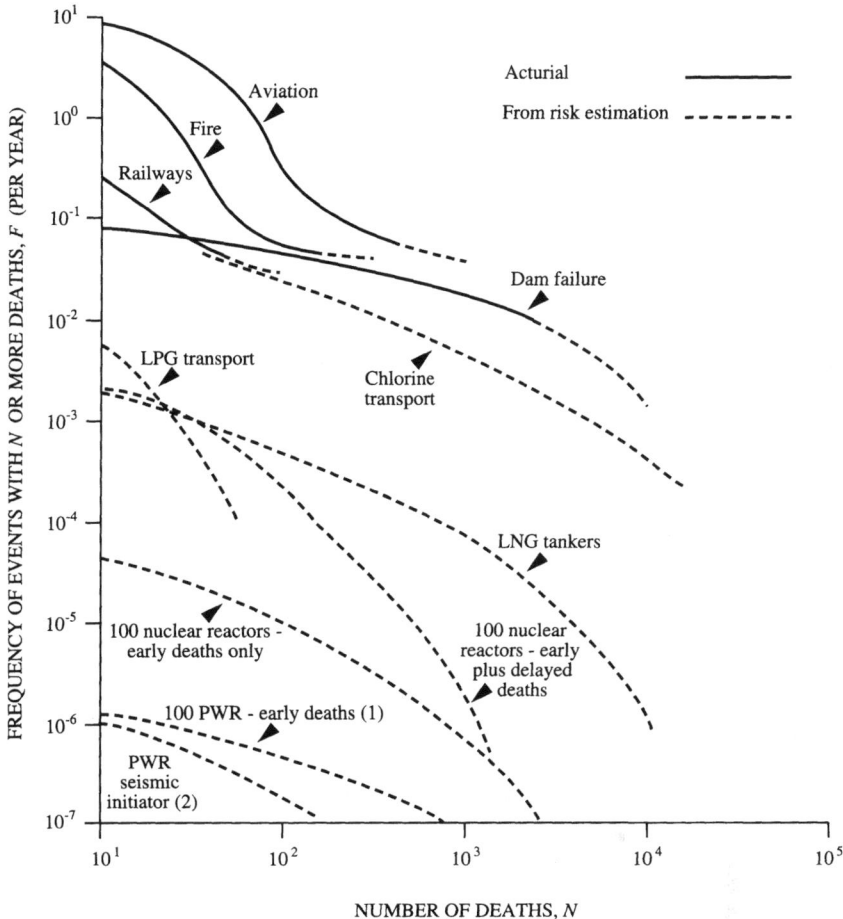

Figure 2.1 Examples of F vs N lines for various man-made hazards. The risk estimates the frequencies are potential events, whereas the actuarial lines are based on actual events. The actuarial lines have extended on the basis of risk assessment. (Sources: Coppola, A. and Hall, R.E., 1981, *A risk comparison, United States Nuclear Regulatory Commission Report (NUREG/CR–1916, BNL–NUREG–51338, R7, RG)*; (1) Gittus, J.H., 1986, Degraded core analysis for the pressurized water reactor, *Sci Publ Affairs*, 2: 121; (2) United States Nuclear Regulatory Commission, 1990, *Analysis of core damage frequency Surry Power Station, Unit 1 — External events, Vol 1, Rev 1, NUREG/CR–4550 Sand 86–2084*.) Reproduced from *Risk Analysis, Perception and Management*, 1992, Figure 1, by permission of The Royal Society.

decision can be made. The matter is complicated firstly by the problems of risk perception already discussed and secondly by the difficulty that most decision-makers have in understanding the basic theory of risk analysis. On top of that, decision-makers must also consider the NIMBY (not in my back yard) syndrome. The basic principle remains that the opinion of the public should underlie the evaluation of risk. Unfortunately there is insufficient public information to allow an understanding of the basis for the regulation of safety.

The politicians do, however, have advisors to help them and to lay down certain ground rules. In the UK this role is undertaken by the Health and Safety Executive (HSE). In fixing acceptability criteria, consideration must be given to the difference between voluntary and imposed risk and to the difference between individual and societal risk. Thus current UK thinking concerning a new or significantly modified potentially hazardous operation looks at how it affects three groups:
- process and other workers directly exposed;
- office and other administration workers on the site;
- the general public.

In general it is accepted in the UK that a new industrial process must not involve risks greater than the industrial average — that is, an overall FAR of greater than 4. A figure of 0.4 is sometimes taken for a single incident. Expressed in another way, the HSE now proposes that the risk of death for workers in a new plant should be less than 1 in 10^{-4} per year. It is accepted, however, that some workers in hazardous areas, who are specially trained and are fully aware of the risks involved, can be subjected to higher risks. In practice such workers usually receive higher pay for the risks involved. The individual risks to office and administration workers who have no such control or special training must be much lower.

In the context of risk the term 'tolerable' is often used rather than 'acceptable'. To quote from an HSE report[5], '"Tolerability" does not mean "acceptability". It refers to the willingness to live with a risk to secure certain benefits and in the confidence that it is being properly controlled. To tolerate a risk does not mean that we regard it as negligible or something that we might ignore, but rather as something we need to keep under review and reduce it still further if and as we can'.

For members of the general public, a figure of 10^{-6} per year is generally considered acceptable[8] for individual risks whilst a risk of greater than 10^{-4} per year would be considered intolerable[5]. As has already been shown there is a relationship between acceptable frequency and number of fatalities. Thus, whilst 10^{-4} per year is tolerable for an individual risk of death due to industrial activity, the figure for an accident likely to cause 100 deaths would be much

16

lower. It is interesting to note that the estimated risk of an aircraft accident in the UK killing more than 500 people is 10^{-3} per year.

2.6   ALARP

The figures quoted in Section 2.2 on page 8 must be used with great caution; whether or not a particular level of risk is tolerable or acceptable depends on the circumstances. Simple limits do not really embrace the question of different perceptions of risk or the benefit, perceived or real, which arises from the particular activity. In the UK the concept of ensuring that detriment (be it to life or the environment) is kept as low as reasonably practicable (ALARP) has a long history. Although not formally defined in legislation, the term has acquired meaning through many interpretations by the courts. In determining what is reasonably practicable, consideration must be given to the benefits of the activity and to the social and economic factors involved. Thus the cost of reducing the risk must not be grossly disproportionate to the risk. The higher the risk the greater the spending to reduce it.

The use of this concept has led to the classification of risk into three regions, as shown in Figure 2.2 on page 18. There will always be a level of risk which is not tolerable under any circumstances, however great the benefits may be. If the risk is considered, or estimated, to be above this level then no further work should be done on the project. There will also be a level below which it is clear, probably intuitively, that the risk is negligible. Under such circumstances no detailed risk analysis is necessary. The region in between these extremes may be termed the ALARP region. In this ALARP region, it is necessary to prove that everything has been done to reduce the risk to as low a level as reasonably practicable.

The point at which the tolerable and negligible lines are drawn is still the subject of debate. The HSE suggests[5] a maximum tolerable individual risk level for workers in any industry of 10^{-3} per year. For the general public the suggested level for large-scale industrial hazards such as major chemical plants is 10^{-4} per year. Because of the public aversion to ionizing radiation, the figures for nuclear installations suggested are 10^{-6} per year for risk of death arising from normal operation and 10^{-7} per year for risk of death arising from accidents.

In assessing the risks associated with new nuclear power installations it has been accepted in the past that the risks, though not negligible, are tolerable because of the benefits of cheap electricity. Similarly, the high risks associated with travel by private car are considered tolerable because of the perceived benefits which private motoring offers the individual.

17

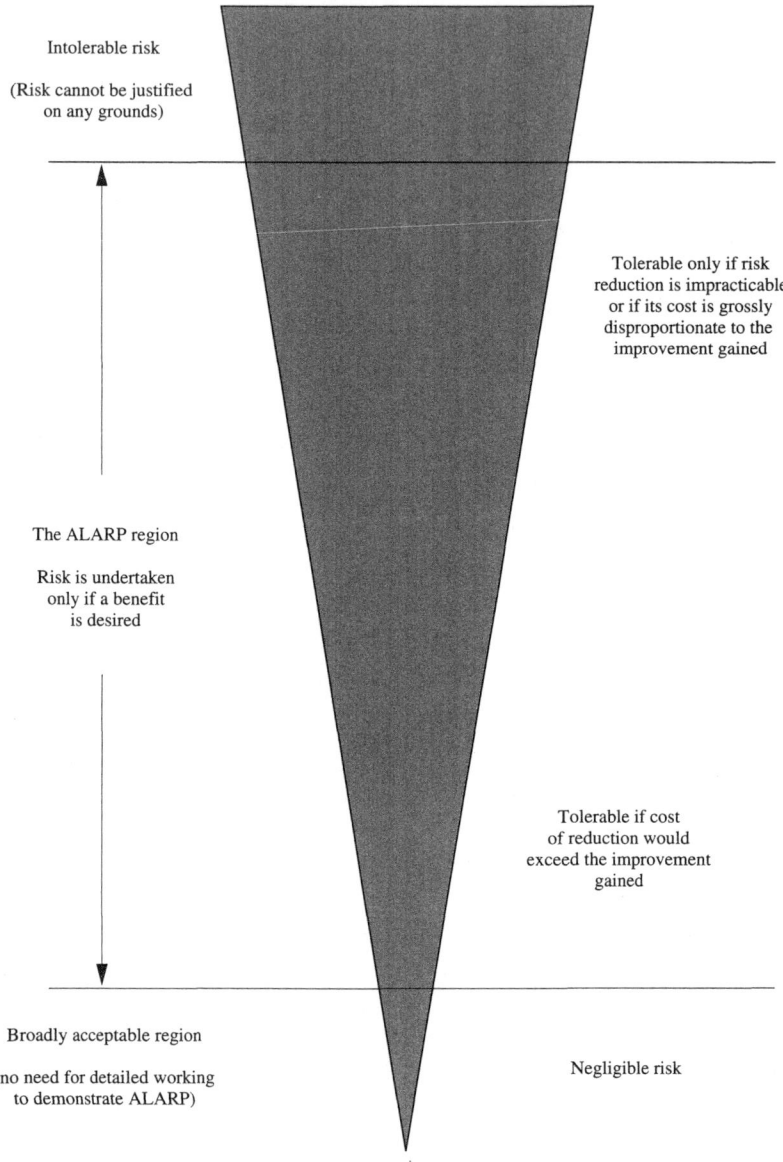

Intolerable risk

(Risk cannot be justified
on any grounds)

Tolerable only if risk
reduction is impracticable
or if its cost is grossly
disproportionate to the
improvement gained

The ALARP region

Risk is undertaken
only if a benefit
is desired

Tolerable if cost
of reduction would
exceed the improvement
gained

Broadly acceptable region

(no need for detailed working
to demonstrate ALARP)

Negligible risk

Figure 2.2   ALARP diagram. (Source: Health and Safety Executive, 1992, *The Tolerability of Risk from Nuclear Power Stations*. Crown copyright is reproduced with the permission of the Controller of HMSO.)

The levels of risks posed by existing chemical plants — such as the Canvey Island complex (after improvements) and the ICI plants in the north of England — are considered to be tolerable because the social and economic consequences of closing down such operations would be too great. Note that the risks imposed by the Canvey Island complex were considered intolerable until certain improvements were made[9]. Before improvements, the individual risk was estimated to be 2570 per million (2.5×10^{-3}) and after improvements, 70 per million (7.0×10^{-5}). Different criteria are likely to be applied to new installations both on existing and particularly on new sites. Similarly, requests for planning permission for new residential developments are likely to be subject to stricter criteria. For significant developments or those near particularly hazardous activities, a limit of 1×10^{-6} is suggested[10].

For societal risk tolerability use can be made of the F/N curve in the setting of criteria. The value of F/N curves has been discussed at length[10] and there is no clear consensus on their use. They only relate to fatalities (or equivalent fatalities) and do not allow for social or economic factors. The aversion factor could be used to weight the curve for larger incidents but this is not a complete solution. The Three Mile Island nuclear plant accident did not cause any deaths or injuries but still generated a great deal of public anxiety and political reaction. The F/N curve can make no allowance for things like the effects of land contamination or the need for mass evacuation. A further factor which complicates the examination of societal risk is that the risk tends to be borne largely by the local population, whereas the benefits may be gained by a much wider population group.

In theory, it should be possible to estimate the F/N curve for a particular activity and superimpose a criteria curve. Figure 2.3 (see page 20), taken from a report on the transport of dangerous substances[11], is an example of a set of criteria curves for a particular operation. There are problems in producing reliable estimate curves for new operations because of the lack of data on high death toll incidents. It is worth noting, however, that this set of curves includes a local scrutiny line as well as the tolerability limit. This allows some degree of consideration of specific local conditions where, under some circumstances, the risk may be tolerable.

The fact remains, however, that the F/N curve has severe limitations as a way of setting criteria. There is no readily deducible and uniformly applicable upper limit of acceptable societal risk[9]. The F/N curve can be used as one item in the decision-making process but because all possible consequences of an accident must be taken into consideration, in the end the decision must be one of informed judgement.

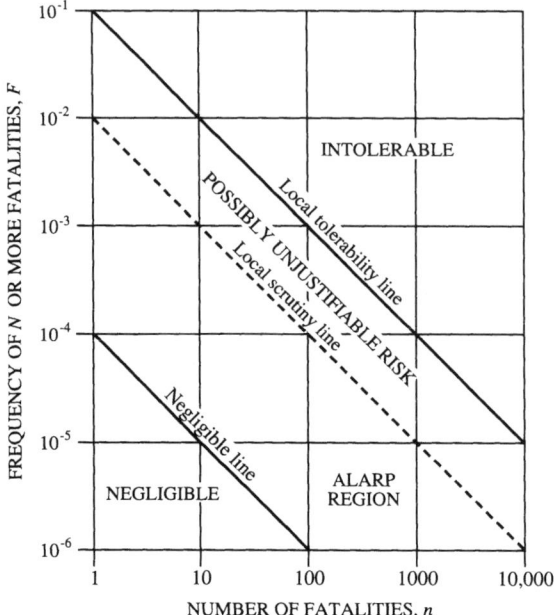

Figure 2.3 F/N curve for transport of dangerous substances[11].

2.7   COST BENEFIT ANALYSIS

Whilst ALARP implies a relationship between risk and benefit, the concept does not in itself impose the requirement for a rigorous cost benefit analysis. In order to do this analysis, more information is needed on both the cost of the detriment and the cost of risk reduction measures. Neither of these costs are easy to obtain.

In order to put a cost on detriment it is first necessary to put a cost on human life. This can be a very emotive subject although the courts and the insurance companies have been doing it for centuries. The Department of Transport has applied cost benefit analysis to road projects for many years; to do this they have had to arrive at costs of both fatal and non-fatal road accidents. The traditional method of assessing the value of a human life is the 'human capital' approach used by the courts.

This approach estimates what a person killed in an accident is likely to have earned for the rest of their life, sometimes with an addition to cover the grief suffered by friends and relations. It gives zero value to the old, the young and the unemployed and does not generally attach any value to housework. Some people find this approach morally repugnant and it does tend to give rather a low value compared with other methods of assessment. In the UK the figure was about £250,000 in 1993.

An alternative approach involves estimating what people would be prepared to spend in order to reduce risk by a certain amount. The problem here is that the risks in general are very low already, so that the reduction in risk is extremely small in absolute terms. Estimates can be obtained by determining what people are actually prepared to spend (revealed preference) and by asking them what they think they would be prepared to spend (expressed preference). The former will tend, in practice, to give a more reliable figure. Again the matter is complicated by perception of risk; for example, people are prepared to travel by car to save relatively small sums of money despite the much higher risk involved compared with rail travel.

Because the amount an individual is prepared to pay for a small reduction in risk is in itself small, a formula can be used to arrive at the cost of a single life saved. If it were found that an individual was prepared to pay £20 per annum for a 10% reduction in the risk of being killed in a road accident, given that the annual risk of death from a road traffic accident in the UK is 10^{-4}, the cost per life saved would be $£20/10^{-5}$ — that is, £2,000,000. This is nearly ten times the 'human capital' cost figure.

The problem can be addressed from the opposite direction: calculate the cost of saving a given number of lives per year and then compare it with a life cost criterion. All calculations are complicated by the need to discount the cost of lives saved in the future by expenditure today. The discount rate used can have a significant effect on the cost benefit of long-term projects such as those concerned with rail safety. The discounted cost of one particular rail safety project (fully automatic train protection which would prevent a driver taking a train past a signal at red under any circumstances) is £60,000,000 per life saved at present day values.

Simple mortality costs cannot allow for the 'fate worse than death' outcome. The courts often give higher awards to this type of injury than to death. The other detriment costs can be estimated in a less emotive manner. Reference 10 gives some guidance on the question of planning gain and risk; provided that the consequence models are reliable, the costs of other effects (that is, no human injury) can be calculated. The question of the total cost of accidents is discussed in more detail in Chapter 3.

Estimations of the costs of risk reduction measures are only straightforward if the measures concerned are purely safety-related. Often, such measures also give operational gains which must be taken into consideration in any cost benefit analysis. Remember also that the costs of safety features included at the design stage will be much lower than if they are added on afterwards;

hence the value of carrying out safety reviews including cost benefit analysis at the earliest practicable stage in a project. Despite its problems, cost benefit analysis has much to offer, particularly when comparing the effects of alternative schemes.

2.8   CONCLUSIONS

The concept of risk is a difficult one for many people but it is essential that all engineers and those responsible for advising decision-makers have a clear grasp of both the qualitative and quantitative aspects of the subject. It is also important that the opinions and expressed views of the general public are understood, in order that safety professionals can distinguish between their own convictions and the fears of the layman. In the past, many problems have been caused by the arrogance of scientists and engineers when a little bit of sympathy and understanding towards the people who seem threatened by some new activity would have been much more productive.

REFERENCES IN CHAPTER 2

1.   *Risk, Analysis, Perception and Management*, 1992 (The Royal Society, London, UK).

2.   Health and Safety Executive, 1978, *Canvey, A Summary of an Investigation of Potential Hazards from Operations in the Canvey Island/Thurrock area* (HMSO, London, UK).

3.   Wynne, B., 1982, *Rationality and Ritual: the Windscale Inquiry and Nuclear Decisions* (British Society for the History of Science, Chalfont St Giles, UK).

4.   Slovic, P., Fishhoff, B. and Lichenstein, S., 1980, Facts and fears, understanding perceived risks, in *Societal Risk Assessment*, Schwing, R.C. and Albers, W.A. (eds) (Plenum Press, New York, USA).

5.   Health and Safety Executive, 1988, *The Tolerability of Risk from Nuclear Power Stations* (HMSO, London, UK).

6.   Hambly, E.C., 1992, Preventing disasters, *Proc Royal Inst London*, 64: 69–92.

7.   Sir Frank Layfield, 1987, *Sizewell B Public Enquiry* (HMSO, London, UK).

8.   Barnes, M., 1990, *Hinkley Point Public Enquiries* (HMSO, London, UK).

9.   Health and Safety Executive, 1989, *Quantified Risk Assessment: its Input to Decision-Making* (HMSO, London, UK).

10.   Health and Safety Executive, 1989, *Risk Criteria for Land Use Planning in the Vicinity of Major Industrial Hazards* (HMSO, London, UK).

11.   Health and Safety Commission, 1991, *Major Hazard Aspects of the Transport of Dangerous Substances* (HMSO, London, UK).

12.   Marion, A., 1986, Evaluating the nation's risk assessors: nuclear power and the value of life, *Public Money*, 6: 41–45.

3. SAFETY IN DESIGN AND OPERATION

Successful risk reduction depends upon both design and organization; in general, good design is only possible with good organization. Safety assurance and quality assurance are thus two sides of the same coin[1]. Human factors must be considered at all stages of safety assurance; the fallibility of the designers can be just as important as that of the operators. This chapter first considers ways of ensuring that risks emanating from the basic design of the plant are kept as low as reasonably practicable and then looks at operational and maintenance aspects. The final section considers the general organizational aspects of health and safety.

3.1   SAFETY IN DESIGN

Safety in process plant starts at the design stage with the first flowsheets; it can be assured in two ways:
- inherent safety;
- safety that is engineered.

INHERENT SAFETY

This is the best way of ensuring safety, because it does not have to rely on the correct functioning of safety devices. Inherent safety includes reducing the inventories of hazardous materials or, if possible, replacing them by less hazardous materials. What you haven't got can't leak![2]. The use of alternative process routes involving lower pressures or more moderate temperatures is another principle of inherent safety. The substitution by more efficient means of earlier process routes which required high energy potentials for product achievement has made a significant contribution towards better inherent safety.

Examples of safety-oriented design philosophy include the use of small continuous reactors for nitration instead of large batch reactors, the substitution of aqueous for oil-based hydraulic fluids and the storage of flammable substances in liquid form in low pressure refrigerated storage rather than as gas at high pressures. Keep stocks of toxic or otherwise dangerous intermediates to a minimum, and eliminate them by the use of continuous processes if possible. The Bhopal accident was largely caused by the storage of excessive amounts of an extremely dangerous intermediate.

In general, design should always ensure that a process fails to a safe or stable condition if power and/or utilities are lost. One essential is to ensure that residual heat can be removed by natural circulation cooling, hence thermo-siphon coolers are preferred to pump circulation systems and natural draught cooling systems to fan-assisted systems.

If possible the design should be such that operator intervention is not needed for at least 30 minutes after an incident. Experience has shown that operators cannot always be relied upon to make the correct decisions under immediate post-accident conditions. This point is discussed in detail in Chapter 9.

The *Approved Code of Practice* for the Management of Health and Safety at Work etc Act Regulations 1992[3] summarizes the concept of safe design very well.

ENGINEERED SAFETY

Properly designed, constructed, operated and maintained equipment will not fail catastrophically provided that its mechanical design conditions are not exceeded, the properties of the materials of construction do not deteriorate and the process conditions remain within specification.

It is thus the first priority of engineered safety to ensure that the basic design is adequate under normal operating conditions. It is then necessary to design protective devices to ensure either that the design conditions cannot be exceeded or that the excessive condition is relieved before it can do any harm.

Examples of such devices include:
- pressure relief devices;
- non-return valves;
- high and low temperature alarms and trips;
- high and low pressure alarms and trips;
- flameproof electrical equipment;
- process interlocks;
- fire detection and fighting systems.

Pressure relief devices are always a last resort; design should be such that the plant is protected by alarms and trips before a dangerous pressure condition is reached. The protection by pressure relief devices discharging to the atmosphere is also being increasingly challenged by the demands of environmental standards.

In protecting against pressure and temperature deviations, consider both high and low conditions. Failures due to low temperature-induced brittle fracture and vessel collapse due to vacuum are not uncommon.

Remember that engineered protective devices can fail. Never place too much reliance on 'bolt on' safety; always give priority to preventing the problem in the first place. The question of reliability of protective devices is discussed in detail in Section 7.2 on Quantitative Risk Assessment, page 112.

3.2   SAFETY ASSURANCE IN DESIGN

The Health and Safety at Work etc Act 1974[4] makes the supplier responsible for the safe operation of plant and equipment supplied. In order to discharge this responsibility the supplier must show that all practicable steps have been taken to ensure a safe design.

Safety in design must be both proactive and reactive. Thus the designer must take positive steps to ensure that the design is safe from the outset. It is not sufficient and certainly not cost effective to carry out the only safety review once the design is complete and then 'bolt on' the safety devices. Safety assurance requires both the identification of potential hazards and an assessment of their frequency and consequences. The techniques used are the subject of the next three chapters. The most common identification technique is the hazard and operability study (Hazop) and the assessment of frequency and consequence is usually termed Quantitative Risk Assessment (QRA).

The key documents in the various safety reviews are the process flow diagrams (PFDs) and the piping and instrumentation diagrams (P&IDs). In some organizations engineering line diagrams (ELDs) are used instead of P&IDs.

Safety assurance must be an integral part of design procedures — that is, part of overall quality assurance. A formal set of review procedures needs to be established to ensure that safety is considered at all points in the design process. The reviews should be proactive and reactive, and carried out at set stages throughout a project. A typical procedure[5] involves carrying out safety reviews at six stages:
- conceptual design;
- completion of flowsheet development;
- basic process design freeze;
- completion of detailed design;
- pre-commissioning;
- completion of first year in operation.

CONCEPTUAL DESIGN
The first safety report is produced at the conceptual design stage. The object of this report is to ensure that all the major safety implications of the process have been considered. If the process is hazardous, make attempts at this stage to find

a less hazardous route or to minimize the risks as far as practicable. Check-lists can be useful to ensure an orderly and structured approach to this initial study. This is the time to consider inherent safety in detail.

The report comprises:

- process description;
- inventory of hazardous materials;
- list of hazardous process operations;
- preliminary assessment of risks and mitigating measures;
- review of previous experience with similar processes or materials.

The report should make all designers and others concerned with the project aware of special precautions needed to ensure a safe design. It is usually prepared pre-sanction or at the tender stage. There have been occasions when a project has been halted at this stage because the risks were considered unacceptable.

FLOWSHEET SAFETY REVIEW

Once the PFDs have been completed it is possible to carry out the initial Hazop review. The Hazop technique is described in full in Chapter 4. The object of this review is to examine the plant section by section using a set of keywords to ensure that the plant will operate safely under normal conditions. The list of keywords could include:

- fire;
- explosion;
- toxic release;
- flammable release;
- pollution.

Although carried out at an early stage of design, this review can show up potential problems and allow changes to be made before the design has reached an advanced state of development.

PROCESS DESIGN FREEZE REVIEW (HAZOP)

Once the process design has reached the point where the P&IDs are complete and can be frozen and 'approved for design', a full-scale Hazop can be carried out. The full Hazop examines the plant line by line in a structured re-examination by applying a series of guide words and deviations. The object is to reveal potential hazards in both normal and abnormal operation. By conducting the study immediately before the 'approved for design' stage, sufficient information is available for a meaningful review but the design is not too far advanced to make changes impossible.

The application of a simplified Hazop at the flowsheet freeze stage of a project is sometimes referred to as Hazop 1 and the more conventional technique as Hazop 2. In some organizations they are termed Hazop 2 and Hazop 3 respectively because they are the second and third safety reviews on the project.

REVIEW AT COMPLETION OF DETAILED DESIGN

Development of the detailed design is bound to result in changes from the plant as subject to the Hazop review. It is thus essential that all subsequent changes to the design are also examined, so that the changes do not prejudice the integrity of the design as originally reviewed. Establish a good project management system to ensure that no changes can be made to the design following the Hazop without the Hazop team being aware of them. Once again, quality assurance and safety assurance go hand in hand.

PRE-COMMISSIONING SAFETY REVIEW

Commissioning is one of the most hazardous parts of any process plant operation. Not only do design errors which escaped previous checks manifest themselves but problems due to construction errors also become obvious. In addition commissioning generates hazards of its own as the plant moves from construction to operating status. It is essential that a formal set of checks be carried out before process fluids are introduced for the first time.

First, check the plant line by line against the P&IDs to ensure that it has been correctly constructed. It must be thoroughly cleaned of all construction debris and a check made to ensure that any instrumentation removed during pressure or electrical testing has been reinstated.

Then carry out a Hazop review to confirm that all outstanding actions have been cleared and that there have been no further changes affecting safety. Pay particular attention to changes made by the commissioning team. At this stage it should be possible to check that the operating instructions are complete and that all safety guards and notices have been provided.

REVIEW AT COMPLETION OF FIRST YEAR IN OPERATION

A further safety review is carried out after the plant has been in operation for about a year. This review includes an examination of the plant records for the year to see if there have been any accidents, or incidents which could have led to an accident. Examine all near-miss incidents to see if there are any intrinsic problems with the plant or the way in which it is operated. Discussion with the operators enables a check that the operating instructions are being followed and, if not, why not. Study any changes made to the plant or to the methods of operation to ensure that they have not reduced the overall safety as originally designed.

It is not unknown for operating personnel to adopt short-cut procedures and manual substitution of automatic control because the control philosophy is not fully understood or training or supervision has fallen short. Such circumstances unwittingly introduce greater risks.

Reputable companies continue to carry out safety reviews or audits on all operating plants at regular intervals by applying the basic methodology described here.

DESIGN SAFETY GUIDELINES

The steps described so far are essentially part of the reactive or checking procedure but they must be backed up by proactive advice as the job proceeds. In today's industrial environment designers have to embrace a wide range of technologies; it is thus not possible to rely on their experience alone to anticipate many aspects of process safety. For this reason it is desirable to issue design safety guidelines and check-lists.

Such guidelines include codes of practice such as those issued under the Health and Safety at Work etc Act 1974 and national standards such as BS and DIN. Many suppliers issue codes of practice for the safe handling of their products. In addition most companies have their own in-house guidelines and standards.

Some guidelines include check-lists which compel designers to examine each area of design against pro-forma questions. The designers must sign that they have considered each point and give reasons if the design does not comply. Provided that the guidelines are well prepared they can save a great deal of re-work and adding-on of safety equipment at a later stage. They have a further advantage that they provide well-documented confirmation that safety checks have been carried out.

Guidelines must always be backed up by proactive safety advice and designers must have someone at hand to whom they can refer for interpretation and advice. All such references and the replies should be documented, partly to assist future designs and partly as evidence that the issues have been considered.

MODEL REVIEW

A further valuable safety review can be carried out when the plant model (plastic or computerized) is complete. Much can be seen on the model which cannot be detected on drawings. A 'walk through' of the model, again using a prepared set of check-lists, can do much to pick up problems which have slipped through earlier nets. Simple matters such as pipes at head or knee height, inaccessible valves, liquid locks in vents and so on are often only detected at this stage. Computer-aided design (CAD) systems often contain packages which can perform many of these checks automatically.

OPERATING INSTRUCTIONS

Operating instructions play a particularly important part in the safe operation of process plant. Ideally they should be written as the design proceeds, not left until the end. They should contain information on all hazards likely to be encountered in the operation of the plant and full details of what to do in the event of abnormal conditions developing.

Instructions should be written in a style understandable by the type of worker likely to operate the plant. Take particular care with operating instructions for plants where the level of supervision will be minimal, plants in remote locations and mobile plant. Instructions written for overseas locations should be very carefully translated and the translation checked by a technically competent local person. For batch plants, consider the inclusion of some form of check-list on which the operator ticks off each sequential operation as it is completed. Thus the exact status of a plant at any time is clear both to the operator and to anyone else concerned.

The production of operating and maintenance manuals is a very costly exercise if done properly, but the consequences of it not being done properly are even greater if there is an accident.

3.3   SAFETY IN OPERATION

The major hazards in the operation of process plant are due to:
- toxic and corrosive chemicals;
- fires and explosions;
- falls and dropped loads;
- mechanical equipment;
- electrical equipment.

Senior management, plant designers and operators are required to be aware of the hazards on their plant and to make every attempt to protect workers from injury. Chapter 8 gives a more comprehensive treatment of the consequences of release of toxic and flammable substances.

CHEMICAL HAZARDS

The Control of Substances Hazardous to Health Regulations[6] (COSHH) are now the main statutory instrument in the UK relating to the dangers from toxic

29

and corrosive chemicals. Plant designs must ensure that such materials are properly contained and assessments are mandatory to evaluate the risks and to show that everything practicable has been done to reduce exposure to a minimum. Where it is necessary for personnel to come into contact with hazardous chemicals at tanker loading stations, sample points and so on, protective clothing must be provided together, if required, with breathing apparatus. Safety showers, eyewash bottles, antidotes and resuscitation apparatus are also required in appropriate locations and the workforce should be trained in their use. Special training is essential for all personnel handling dangerous materials and medical surveillance may be required in some circumstances.

Always regard the use of protective equipment — particularly breathing apparatus — as the last resort. The plant should be designed in such a way that noxious substances are fully contained or the concentration controlled — for example, by adequate ventilation.

Full details of maximum permitted occupational exposure limits (OELs) are available in a number of publications including HSE's *Guidance Note EH40* [7]. For fuller details on the toxic properties of materials see *Sax's Dangerous Properties of Industrial Materials* [8] which is generally regarded as the definitive work in this area.

FIRES AND EXPLOSIONS

The chemical industry handles a wide variety of flammable materials, so fires and explosions are common hazards. Gas/air mixtures only explode if their composition is within a fixed range known as the lower and upper explosion limits (LEL and UEL). Chapter 8 provides a comprehensive list of explosion limits and other relevant data.

Mixtures within the flammable limits can be ignited either by a source of ignition of sufficient energy or, if the temperature is high enough, by spontaneous ignition. Liquids can only be ignited if their temperature is above a limit known as the flashpoint. Gases do not have flashpoints and can be ignited at any temperature. Liquids with flashpoints below 28°C are classed as highly flammable and demand special precautions.

Fires and explosions can be prevented either by ensuring that flammable limits are not reached, flashpoints are not exceeded (liquids) or that sources of ignition are excluded. In practice it is desirable not to exceed 25% of LEL if possible. Flammable atmospheres can be avoided by ensuring that fuel lines and tanks are pressurized so that the flammable material leaks out rather than air leaking in. Similarly good ventilation of vessels and plant areas handling flammable gases can maintain safe working conditions. For this reason process plant is best situated in the open so that flammable material is less likely to accumulate.

Any combustible dust will explode given the right conditions. Such explosions used to be very common in flour mills, coal mines and wood-working shops producing large amounts of sawdust. Dust explosions are best prevented by good housekeeping — that is, by keeping the concentrations of dust down and perhaps keeping the dust damp. Inerting by dilution with non-combustible dust is another effective technique, frequently used in coal mines.

IGNITION

If a non-flammable atmosphere cannot be guaranteed, then the only alternative is to remove or suppress all sources of ignition. This can be effected by the use of flameproof electrical equipment and excluding all other sources of ignition by administrative controls (no matches, no hot work).

Flameproof electrical equipment is very expensive and can be difficult to maintain. It is therefore necessary to restrict its use to those areas where it is absolutely essential. This is normally done by means of a system of zoning in which the plant is divided into zones depending on the risk of presence of flammable atmospheres. The guide for zoning used in the UK is BS 5345[9]. This standard gives three zones as follows:

- Zone 0 — Explosive gas/air mixture continuously present or present for long periods;
- Zone 1 — Explosive gas/air mixture likely to occur in normal operations;
- Zone 2 — Explosive gas/air mixture not likely to occur in normal operations and, if it occurs, only existing for a short time.

A non-hazardous area is defined as one in which an explosive gas/air mixture is not expected to be present in such quantities as to require special precautions for the construction and use of electrical apparatus.

Different types of flameproof electrical equipment are required in each zone. The problem of deciding zone boundaries can be very complex and Cox et al[10] gives more guidance on this matter than can be obtained by reference to the British Standard. The production of the zoning drawings showing the boundaries between the zones should be done at an early stage in the project and the zoning drawings must be available for the Hazop study.

STATIC ELECTRICITY

Particular attention must be paid to the problems of static electricity — the cause of many explosions in industry. Static electricity can be generated, for example, by conveyor belts, the pumping of liquids from nozzles into free air space, and by the use of certain floor coverings. The risk is only serious if the relative humidity is below about 60% and hence is a greater problem in dry countries and in air-conditioned buildings. Precautions include the use of treated materials to

31

prevent static build-up, ensuring that tank filling connections discharge below liquid level and good earthing. It is sometimes necessary to earth bond every flange connection and it is particularly important that tankers are earthed before the transfer of liquid.

EXPLOSION VENTING

If explosions are a possibility, much can be done to mitigate the potential damage by ensuring the venting of the gases generated to locations where they can do least harm. Buildings may be fitted with very light roof or wall panels designed to fail in the event of an explosion. Bursting discs can vent gases from vessels but they must be directed to a safe outlet such as a dump vessel. In some cases it may be possible to design process vessels so that they can contain the maximum pressure rise generated by an explosion. In practice this may mean strengthening to eight or nine times the normal design pressure. Chapter 8 deals with the question of pressure rise due to explosions.

Explosions in pipes and ducts can be particularly dangerous as they can transfer effects to other parts of the plant. Explosion relief and perhaps flame traps should be provided at intervals to prevent this possibility. As with bursting discs they must be located so that they cannot harm personnel if activated. Plants with a very large explosion risk — for example, explosives manufacturing facilities — are usually designed so that the buildings are separated by a safe distance and surrounded by earth mounds. Any explosion will then go upwards rather then affect other plants in the area. In addition there is usually a limit on the number of people allowed in a building at any particular time, thus limiting the death toll in the event of an accident.

FIRE-FIGHTING

Due to the wide differences in the behaviour of materials on fire, the provision of fire detection and fighting is a specialist subject beyond the scope of this book. All factories have to be provided with suitable apparatus to detect and fight fires. Fire-fighting appliances must be clearly labelled and staff instructed in their use. Particular attention must be paid to the location and selection of the correct extinguisher for the types of fire concerned. It is also essential that the contents of buildings are clearly identified, so that fire-fighters know exactly what the risks are and what methods to use. There have been many instances of oxidizing agents stored in the same area as highly combustible organics, making the spread of fires much more rapid and extinguishing much more difficult.

The requirements of the Building Regulations[11] must be observed on matters of fire safety, particularly as regards the provision of escape routes. There is often a conflict between security and fire exits. This may be resolved by the use of special quick-release, alarmed locks on safety doors.

One problem which is currently attracting attention is the means of safe disposal of fire-fighting water. It has long been known that fire-fighting water often causes more damage than the fire itself. Only recently, however, has the environmental damage caused by fire-fighting water been recognized. Fires at the Sandoz plant in Switzerland and the Allied Colloids plant in the UK both caused significant damage to river life and water supplies due to the pollutants released by the fire being washed over the plant boundaries into nearby water-courses. Consideration should be given to impounding and possibly reusing fire-fighting water. In some cases other means of fire-fighting should be used. Unfortunately, the best alternative — the Halon group of gases — is out of favour because CFCs are environmentally suspect. In some cases it may be better simply to let the fire burn itself out, particularly if it poses no problems to adjacent plant. It is then necessary to concentrate all attention on stopping the spread by, say, keeping adjacent areas cool and providing a fire break. There may be a conflict between accepting the atmospheric pollution caused by letting the fire burn out and the water-borne pollution caused by fighting it.

OTHER HAZARDS

Non-process hazards account for more than 70% of all accidents in process plant. The precautions required are the same as in other industries and particular attention must be paid to matters such as machine guards, safe access and electrical safety.

3.4   MAINTENANCE

Many of the worst accidents in the process industries are the result of bad maintenance practice. Piper Alpha and Flixborough are two particularly severe examples. The consequences of accidents which are maintenance-related generally fall into two categories:

- accidents to maintenance personnel;
- accidents resulting from improper maintenance procedures.

The former are very common and can best be avoided by rigorous permit-to-work procedures. The latter require a detailed examination of all maintenance procedures before starting work.

PERMITS-TO-WORK

A permit-to-work is a document signed by a responsible individual indicating that the required steps have been taken to render safe the area or job and its related human interface. Figure 3.1 on pages 34–35 is an example of a permit-to-work.

PERMIT-TO-WORK		
Plant:	**Section:**	
Details of work:		
Withdrawal from service:	The plant has been withdrawn from service Signed: Date Time:	
Isolation:	The equipment has been isolated from all sources of dangerous gas and fume by means of: Signed: Date: Time:	
	The equipment has been isolated from all sources of electrical power Signed: Date: Time:	
	All drives and other sources of mechanical energy have been isolated Signed: Date: Time	
	The equipment has been isolated from all sources of heat and cold Signed: Date: Time	

Figure 3.1   Example of a permit-to-work.

	Valid until (date and time):	**No:**
Testing:	Gas tests have been carried out and the results are as follows: Signed: Date: Time:	
Certification:	It is safe for the work to start with/without breathing apparatus. The following protective gear is required: Signed: Date: Time:	
Acceptance:	I have read and understand this permit and will undertake the work in accordance with the conditions in it. Signed: Date: Time:	
Completion:	The work is now complete and all persons under my control, materials and equipment have been withdrawn. Signed: Date: Time	
Cancellation:	This permit is now cancelled and the plant may be returned to service. Signed: Date: Time	

Before issuing a permit-to-work, the following requirements must be satisfied:

- electrical power isolated;
- mechanical drives isolated;
- sources of heat or cold isolated;
- sources of flammable and toxic gases isolated;
- working area tested for breathable atmosphere.

The definition of isolation is very important. Electrical power is isolated by locked and tagged switches at the distribution board. Mechanical drives and suspended loads are physically secured with scotches or props, again locked in place. Sources of dangerous material are isolated by spades, double block and bleed valves or removal of a section of the feed pipe. A single valve, even if locked, is not sufficient. If the work involves entry into a confined space, including open pits, the space is first tested for flammable and toxic gases and for a breathable atmosphere. The appropriate detectors remain in use throughout the work period. In certain situations the permit may demand the use of protective clothing or breathing apparatus, but breathing apparatus should always be considered as a last resort.

Remember that as many people die by asphyxiation as from toxic gases. Nitrogen is often used to purge toxic or flammable gases and care must be taken to ensure that the nitrogen has then been purged out with air. The increasing use of argon for welding must also be considered as a source of hazard. There have been a number of accidents caused by the failure of temporary pipe dams, allowing argon to escape from the weld zone to other parts of the plant.

Permit procedures vary to some extent but the following points are significant in all cases:

- no work to be carried out without a permit;
- permits only issued and cancelled by a responsible person;
- copy of the permit to be carried by the worker;
- copy of the permit to be held in the control room;
- key to locks to be held by the worker;
- all affected controls to be tagged and locked;
- permits to have a defined time limit.

The procedure for issue and cancellation is critical to safe operation. The permit-to-work must not be issued until the authorizer has personally confirmed that isolation and testing are complete and the isolation must not be removed until the permit is cancelled. Particular attention is required at shift change-over to ensure that the new shift is aware of all permits in force. The permit system should be audited regularly to check that it is being correctly applied. There was a permit system in force on Piper Alpha at the time of the accident but it was not being operated correctly. The senior onshore management were living with a false sense of confidence that all was well offshore, when in fact a serious accident was waiting to happen.

MODIFICATION PROCEDURES
Whilst Piper Alpha was caused largely by a faulty permit system, Flixborough was the result of a badly planned maintenance procedure.

If maintenance involves operating the plant with some form of temporary equipment, a full safety review of the proposed modification must be carried out. This can be done using a system of documentation often called a 'plant modification procedure' which details the changes and the reasons for them. Figure 3.2 on page 38 shows a typical example. The modification can then be subjected to a form of Hazop before permission is given to proceed. Very often the simple initial Hazop method (see page 26) using the basic guide words is all that is necessary. The safety review is applied to the procedures for carrying out the modification as well as the modification itself; such procedures should be followed even for very short-term modifications if the plant is to be operated.

Had the proposed modifications at Flixborough been properly documented and subjected to even the most basic of safety reviews by competent persons, then the accident could almost certainly have been avoided[12].

3.5   ORGANIZING FOR SAFETY

High standards of safety, both at the design stage and during operation, will only be achieved if the organizations concerned have a good safety culture. To paraphrase HM Chief Inspector of Factories, 'there are no short cuts to good health and safety, the starting point must be a genuine commitment of top management and this commitment must permeate from the top to the bottom of the organization'[1]. A good safety culture ensures that both the spirit and the letter of the law are fulfilled.

The key elements of successful health and safety management can be summarized as follows[1]:

- policy;
- organizing;
- planning;
- measuring performance;
- auditing and reviewing performance.

A good health and safety policy is always cost effective; most organizations grossly underestimate the cost of accidents[13], often by an order of magnitude. Thus the policy should satisfy shareholders as well as employees.

PLANT MODIFICATION PROCEDURE						
Plant:		**Section:**		**Date:**	**No:**	
Drawing number(s):						
Details of work:						
Justification (reasons and costs):						
Consequences if not implemented:						
Programme:						
	Signed	Date			Signed	Date
Prepared by			Project approval			
Process approval			Hazop yes/no			
Engineering approval			Safety approval			
Operations approval			Authorized			

Figure 3.2   Example of a plant modification procedure.

The organization should be such that the attitude to safety is highly visible and shared at all levels within the company. Active participation is encouraged to promote the objectives of not just preventing accidents and industrial illness but motivating and empowering everyone to work safely.

Such standards can only be achieved by good planning; risks are identified, methods of eliminating them developed and the progress towards their elimination planned and monitored. Failure to meet the standards must be investigated with a view to preventing recurrence rather than to apportioning blame.

All near-misses and other incidents revealing health and safety implications are studied to see if anything can be learned to help in the future.

The health and safety record and the procedures involved in maintaining it should be the subject of regular reviews and independent audits. These form the basis for self regulation and for securing compliance with Sections 2 to 6 of the Health and Safety at Work etc Act 1974[4].

Accident rates can be greatly reduced by careful personnel selection, training and motivation. A safety culture, once established, must be maintained, and any tendency to careless practices stamped out at once. Experience shows that 80% of accidents tend to happen to 20% of the workforce — the young and the old being particularly vulnerable. So it is essential that young people are properly trained and older people removed from jobs if they are no longer up to them. All training must be backed up by practice sessions and regular exercises to ensure that people know what to do in the event of an incident. Retraining is very important even if a job changes only by a small amount. Experience has shown that many accidents are caused by operators not fully appreciating the significance of small but nevertheless important changes.

For a more detailed discussion of this topic the reader is referred to the IChemE publication *Developing Effective Safety Systems*, by Ian G. Wallace.

3.6 ACCIDENT INVESTIGATION AND REPORTING

All accidents must be investigated and reported, not to apportion blame but to ensure they are not repeated. This is also important for legal and insurance reasons in case there are delayed claims for compensation. Of particular importance is the reporting of minor events and near-misses, which can have significant safety implications and lessons for the future. A good system of reporting is proactive and reactive, whereas most tend to be purely reactive.

The steps in both proactive and reactive monitoring can be summarized as follows[1]:

- assess action necessary to deal with the immediate risks;
- assess the level and nature of investigation required;
- investigate;
- analyse results;
- review.

The level of investigation in terms of detail and the status of investigators is commensurate with the actual or potential harm. It includes organizational aspects as well as considering the actual task being done and the person doing it.

In analysing and reviewing results an attempt should be made to identify any common causes which may reveal weaknesses in the overall safety culture. Some form of database or a simple coding system may help in this respect.

A typical accident report includes the following information[1]:

- details of injured persons;
- details of circumstances and conditions at time of accident;
- details of event — including initiating events, the direct cause, any underlying causes, organizational problems and so on;
- outcome of accident, injuries to workers, damage to plant, emissions to the environment, third party injuries and damage, details of immediate response;
- was the event preventable and if so, how?
- what was the worst that could have happened?
- what prevented the worst from happening?
- how often could it occur?

Whilst near-miss reporting should be similar to actual accident reporting, it can be complicated by the question of blame. Whilst it should never be the object of accident reporting to find scapegoats, human nature being what it is, this will always be a factor. Very often in a near-miss situation, the person who may have made a mistake feels lucky (on this occasion) to have 'got away with it'. There will therefore be a reluctance to say anything about it and the threat of prosecution can make matters much worse. Some organizations attempt to operate a 'no blame' culture to encourage reporting — for example, the UK Civil Aviation Authority has an anonymous reporting system for near-misses in aircraft movements. Following the prosecution of a pilot for a serious misjudgement of his approach to London Heathrow, however, there was a significant fall in the number of reports made. The threat of prosecution also delays and in some cases seriously undermines accident investigations. Following the imprisonment of a train driver in the UK after an accident there has been a noticeable reluctance of train crews to participate in accident enquiries in case what they say is used in evidence against them.

3.7   CONCLUSIONS

Effective safety at all stages of a project — from inception to final demolition — can only be achieved if there is a commitment at all levels. The senior management must see health and safety as being just as important as profitability and they must make certain that all their workers are aware of this fact. The right organization and procedures must be established to ensure that both proactive and reactive safety checks and reviews are carried out at all stages of design, construction, commissioning and operation; also that these procedures are regularly reviewed and audited. There must be a commitment to continually improve the health and safety record and culture of the organization. If an accident does occur the sole emphasis, once the immediate consequences have been mitigated, must be to ensure that it does not happen again.

A well-managed company is almost invariably not only a profitable company but a safe company.

REFERENCES IN CHAPTER 3

1. Health and Safety Executive, 1991, *Successful Health and Safety Management, HS(G)65* (HMSO, London, UK).
2. Kletz, T.A., 1991, *Plant Design for Safety — A User-Friendly Approach*, 2nd edition (Taylor & Francis).
3. Management of the Health and Safety at Work etc Act Regulations 1992, *Approved Code of Practice* (HMSO, UK).
4. Health and Safety at Work etc Act 1974 (HMSO, London, UK).
5. Turney, R.D., 1990, Designing plants for 1990 and beyond: Procedures for the control of safety, health and environmental hazards in the design of chemical plant, *Trans IChemE*, 68 (B1): 12–16.
6. Control of Substances Hazardous to Health Regulations, 1988 (HMSO, London, UK).
7. Health and Safety Executive, 1984, *Occupational Exposure Limits, Guidance Note EH40* (HSE, Bootle, UK).
8. *Sax's Dangerous Properties of Industrial Materials*, 6th edition, 1984 (Van Nostrand Reinhold, New York, USA).
9. BS 5345: Part 2: 1983, British Standard code of practice for selection, installation and maintenance of electrical apparatus for use in potentially explosive atmospheres (other than mining applications or explosive processing and manufacture). Part 2: Classification of hazardous areas (BSI, UK).
10. Cox, A.W., Lees, F.P. and Ang, M.L., 1990, *Classification of Hazardous Locations* (IChemE, Rugby, UK).
11. Building Regulations 1991 (HMSO, London, UK).
12. Kletz, T., 1988, *Learning from Accidents in Industry* (Butterworths, London, UK).
13. Moore, K., 1993, Counting the costs of accidents, *Competitive Edge*, Issue 1, Summer 1993 (AEA Technology, UK).

FURTHER READING

1. Wallace, I.G., 1995, *Developing Effective Safety Systems* (IChemE, Rugby, UK).

4.   HAZOP

4.1   INTRODUCTION TO HAZOP

The term 'Hazop' originated in ICI and first appeared in the literature in the early 1970s[1]. One of the first books to describe its use was published in 1977[2]. Hazop is now used universally to define the application of a formal, systematic, critical, rigorous examination to the process and engineering intentions of new and existing facilities to assess the hazard potential of mal-operation or mal-function of individual items of equipment and the consequential effects on the facility as a whole. Though developed primarily as a safety review, the word actually comes from hazard and operability review and can thus also be valuable in detecting operability problems. It is often found that an examination of an operability problem will reveal a potential hazard.

OBJECTIVE OF HAZOP

The objective of Hazop is to stimulate the imagination of a review team, including designers and operators, in a systematic way so that they can identify potential hazards in a design. Though originally developed for continuous plant, the technique can, with some modifications, be applied to batch processes, company operating procedures and items of equipment. It is equally applicable to old and new projects.

The term Hazop is sometimes applied to a complete range of safety reviews, though the technique as originally published was intended to be carried out at a specific stage in the development of a project. In a number of organizations this is now termed HS3 and is performed at the 'process design freeze' stage[3]. This matter has already been discussed in Chapter 3.

Hazop generates a record and provides proof that a recognized form of hazard analysis has been applied to the project. For many operations, a Hazop study provides all the safety assurance that is necessary to satisfy the regulatory authorities.

Hazop provides an understanding of the causes and consequences of deviations from expected behaviour and facilitates decision-making on actions needed to eliminate or reduce the risks. It must be remembered, however, that Hazop is an identifying technique and is not intended as a means of solving problems. Although Hazop is essentially a qualitative technique, it can be used to identify areas which must be subjected to comprehensive quantitative analysis. It can also be semi-quantified by the use of simple ranking systems.

The Hazop study is not intended as a substitute for good initial design and the proper application of safety codes. It must not be seen purely as a design checking function; normal design quality assurance should be applied irrespective of whether or not the project is being subject to Hazop. The strength of the Hazop is that it examines the system as a whole as a team effort, whereas individual designers normally only check their own areas of interest.

The application of Hazop at the correct stage in a project means that problems are identified and can be rectified during detailed design. This results in substantial savings; changes once a plant is built are very expensive compared with changes at the design stage.

The Hazop can also provide a considerable amount of useful material for inclusion in the plant operating instructions, thus resulting in better informed operations personnel and safer operation.

HAZOP STUDY TEAM

The Hazop team normally comprises between four and eight members, each of whom can provide knowledge and experience appropriate to the project to be studied. The team needs to be small enough to be efficient and allow each member to make a contribution, whilst containing sufficient skills and experience to cover the area of study comprehensively. It is unlikely that a team of less than four can provide the necessary degree of expertise and experience, whilst teams of more than eight people can become difficult to manage. All members of the team must have enough technical knowledge and authority to make decisions within their own orbit of responsibility.

Two types of person are required in a Hazop team:
• those with detailed technical knowledge of the process;
• those with knowledge and experience of the Hazop technique and the ability to chair and report upon technical meetings.

A typical team for a new project comprises:
• chairman or team leader;
• secretary;
• process design engineer;
• control engineer;
• operations specialist;
• project engineer.

Other specialists may be consulted or be available for specific points.

The chairman is selected for his or her ability to effectively lead the study. He or she should have sufficient seniority to give the study recommendations the proper level of authority, and ideally be independent of the project.

In order that the results of the study can be properly recorded, the secretary should have a technical appreciation of the project and be familiar with the Hazop technique. It is a good training ground for future Hazop chairmen.

The technical members of the team are usually part of the project design team, so that they can answer questions directly. Input from operations or commissioning personnel is essential if all operability-related problems are to be identified.

The presence of the project manager or representative can be useful as it shows the importance of the Hazop study in the overall project plan and can facilitate a rapid response to actions/recommendations. Some organizations require the presence of a safety specialist, though this role may well be filled by the chairman.

HAZOP DOCUMENTATION REQUIREMENTS

The principal document for a Hazop is the P&ID but the following documents are essential at the start of the study:
- process and instrumentation diagrams (P&IDs);
- process flow diagrams (PFDs);
- general arrangement drawings;
- relief/venting philosophy;
- chemical hazard data;
- piping specifications;
- process data sheets;
- previous safety reports.
    Other desirable documents include:
- operation and maintenance instructions;
- safety procedure documents;
- vendor package information;
- piping isometrics.

The needs for each study are different and the manner in which the information is assembled differs from project to project. However, sufficient documentation should always be available to indicate clearly the design intent and to provide details of process parameters. Additional information is needed for batch processes as discussed later.

4.2 THE BASIC CONCEPT OF HAZOP

The concept of Hazop involves the splitting up of the plant into sections and the systematic application of a series of questions to each section. The study team discovers how deviations from the design intent can occur and can decide the consequences of the deviations from the points of view of hazard and operability. The following terms[2] are used as a basis for all Hazop studies:

Design intent — the way in which the plant is intended to operate.

Deviation — any perceived deviations in operation from the design intent.

Cause — the causes of the perceived deviations.

Consequence — the consequences of the perceived deviations.

Safeguards — existing provisions to mitigate the likelihood or consequences of the perceived deviations and to inform operators of their occurrence.

Actions — the recommendations or requests for information made by the study team in order to improve the safety and/or operability of the plant.

Guide words — simple words used to qualify the intent and hence discover deviations.

Parameters — basic process requirements such as 'flow', 'temperature', 'pressure' and so on.

DIVISION INTO SECTIONS

In order to proceed in a logical and efficient manner the P&ID is first divided into sections. Too many small sections will increase the workload and lead to a great deal of duplication; however, the division of the P&ID into too few very large sections can result in important deviations and consequences being missed. In general it is better to err on the side of having too many sections than too few. The most important point in sectioning is that the guide words must apply uniformly throughout every part of the section. Factors to be considered when sectioning the plant include:

- purpose/function of the section;
- material (volume or mass) in the section;
- material process/state considerations;
- reasonable isolation/terminal points;
- consistency of approach.

      The general guidelines to be followed can be summarized as follows:

- define each major process component as a section. Usually anything assigned an equipment number is considered a major component;

45

- define one line section between each major component;
- define additional line sections for each branch off the main process flow;
- define a process section at each connection to existing equipment.

USE OF GUIDE WORDS

The questions are formulated using a number of guide words to ensure a consistent and structured approach. The application of an accepted set of guide words ensures that every conceivable deviation is considered. The guide words are normally applied in conjunction with a series of process parameters to arrive at a meaningful deviation. The Hazop study could perhaps be considered as a structured or guided brainstorming session.

The development of meaningful deviations from the guide words depends on the nature of the process being studied. Two approaches are possible:
(1) Select a guide word — say, NONE — and apply it in turn to a number of process parameters — for example, FLOW, TEMPERATURE, PRESSURE — to produce such meaningful deviations as NO FLOW, NO PRESSURE and so on.
(2) Select a process parameter — say, FLOW — and apply the guide words in turn to produce such meaningful deviations as NO FLOW, MORE FLOW and so on.

Both techniques should give the same results but many practitioners now consider that the second approach is more logical.

Tables 4.1 and 4.2 list typical guide words, parameters, deviations and possible causes. Table 4.1 gives parameter-related guide words and Table 4.2 (see page 48) gives general guide words. Such general guide words are often applied to larger sections of the plant than the parameter-related guide words. Many companies have produced their own lists for specific industries.

QUANTIFICATION

Whilst the Hazop technique was originally conceived as a purely qualitative review, it can be quantified to some extent by the use of a simple ranking for likelihood (or frequency or probability) and severity (or consequence). A typical ranking is as follows:

Likelihood ranking
1   Very unlikely
2   Unlikely
3   Has happened once
4   Has happened a few times
5   Happens quite often

TABLE 4.1
Process parameter-related guide words

NO FLOW	Wrong routing — blockage — incorrect slip plate — incorrectly fitted NRV — burst pipe — large leak — equipment failure — incorrect pressure differential — isolation in error — no material available — vapour lock.
LESS FLOW	Line restriction — filter blockage — defective pumps — fouling — density or viscosity problems — incorrect specification of process fluid.
REVERSE FLOW	Defective NRV — syphon effect — incorrect differential pressure — two way flow — emergency venting — incorrect routing.
MORE FLOW	Increased pumping capacity — increased suction pressure — reduced delivery head — greater fluid density — exchanger tube leaks — restriction orifice plates deleted — cross connection of systems — control surging — valve(s) failed open.
LESS PRESSURE	Vacuum condition — condensation — gas dissolving in liquid — restricted pump or compressor suction line — undetected leakage — vessel drainage.
MORE PRESSURE	Surge problems — leakage from interconnected hp system — gas breakthrough — isolation procedures for relief valves defective — thermal overpressure — positive displacement pumps — failed open PCVs — uncontrolled reaction.
LESS TEMPERATURE	Ambient conditions — reducing pressure — fouled or failed exchanger tubes — loss of heating.
MORE TEMPERATURE	Ambient conditions — fouled or failed exchanger tubes — fire situation — cooling water failure — defective control — fired heater control failure — internal fires — reaction control failures.
LESS VISCOSITY	Incorrect material specification or temperature.
MORE VISCOSITY	Incorrect material specification or temperature.
COMPOSITION CHANGE	Leaking isolation valves — leaking exchanger tubes — phase change — incorrect feedstock/specification — inadequate quality control — process control.
MORE THAN	Contamination leaking exchanger tubes or isolation valves — incorrect operation of system — interconnected systems — effect of corrosion — wrong additives — ingress of air — impurities — extra phases.

TABLE 4.2
General guide words

OTHER ACTIVITIES	Start-up and shutdown of plant — testing and inspection — commissioning — decommissioning — demolition — weather — seismic — physical impact — containment loss and consequences — domino effects — purging — washing out — toxicity.
RELIEF	Relief philosophy — type of relief device and reliability — relief valve discharge location — pollution implications.
CONTROL	Control philosophy — location of instruments — response time — set points of alarms and trips — time available for operator intervention — alarm and trip testing — fire protection — electronic trip/control amplifiers — panel arrangement and location — auto/manual facility human error.
SAMPLING	Sampling procedure — time for analysis result — calibration of automatic samplers/reliability — accuracy of representative sample.
CORROSION/ EROSION	Cathodic protection arrangements — internal/external corrosion protection — engineering specifications — zinc embrittlement — stress corrosion cracking — fluid velocities — riser splash zones.
SERVICE FAILURE	Instrument air/steam/nitrogen/cooling water — hydraulic power — electric power — telecommunications — heating and ventilating systems — computers.
MAINTENANCE	Isolation — drainage — purging — cleaning — drying — slip plates — access — rescue plan — training — pressure testing — work permit system — condition monitoring — catalyst change and activation.
STATIC	Earthing arrangements — insulated vessels/equipment — low conductance fluids — splash filling of vessels — insulated strainers and valve components — dust generation and handling — hoses.
SPARE EQUIPMENT	Installed/non-installed spare equipment — availability — modified specifications — storage of spares.
SAFETY	Fire and gas detection system/alarms — emergency shutdown arrangements — fire-fighting response time — emergency and major emergency training — contingency plans — TLVs and methods of detection — noise levels — security arrangements — knowledge of hazards of process materials — first aid/medical resources — effluent disposal — hazards created by others (adjacent storage areas/process plant, etc) — testing of emergency equipment — compliance with local/national regulations — environmental considerations.

Severity ranking

1   Pollution
2   Minor injury/damage
3   Serious injury/damage
4   Fatalities/major damage
5   Loss of installation

The two factors can be combined in a matrix to give an overall ranking which can then be used to prioritize actions. The growing use of computers has tended to encourage the use of such ranking schemes.

4.3   CONDUCTING A HAZOP STUDY

In order to make the Hazop technique work effectively, it is necessary to institute a formal procedure including the following headings:

- define objectives and scope;
- prepare for the study;
- carry out the study;
- record the results;
- follow up.

DEFINE OBJECTIVES AND SCOPE
It is essential that the objectives and scope are clearly understood by all concerned, and documented and agreed before the start of the study. The definition should include, but not necessarily be limited to, the following:

- study terminal points, best defined in terms of P&IDs;
- design status at time of study, defined in terms of P&ID revision status;
- extent to which effects on and by adjacent plant should be considered;
- study programme including action reply and final reporting dates;
- links with studies being conducted on adjacent or related plants.

PREPARE FOR THE STUDY
The timing of the study is absolutely crucial to its success. The timetable depends upon:

- project programme dates;
- availability of documentation;
- availability of personnel.

As stated earlier, the ideal time to carry out the Hazop is at the process design freeze. Many companies use the completed Hazop as a signal that the P&IDs are 'approved for design'. At this stage the process design should be sufficiently advanced for most of the required information to be available but not sufficiently advanced for alterations to be too costly.

Further Hazop reviews are needed at the next stage of design freeze to ensure that changes in the development of the detailed design have not negated the original Hazop.

The Hazop must be an integral part of project planning; it is a project time and a man-hour consuming exercise and further delays are inevitable whilst actions are being cleared. Allow at least one day per P&ID and about three weeks for action clearance.

There is no point in starting a study until all the documents are available. Commencing with insufficient information delays progress and inevitably affects the credibility of output. The chairman or co-ordinator must ensure that all necessary documentation is available at least one week before the start of the study. All documents should have been checked and be at the required issue status for the study. It is normal for A0 size prints of the P&IDs to be available as the Hazop record copies and A3 size copies to be available for other team members. A0 prints of the PFDs are displayed in the meeting room. Single copies of the other documents should suffice.

Because of the importance of the Hazop in any project programme, the project manager must ensure that all necessary personnel are available. Since the study can take place over an extended period, it is essential that the long-term commitment is understood at the start of the study and allowed for in the project programme. Team members are not changed during the course of a study unless it is absolutely essential. If substitutes are needed there should be a handover meeting with both original and substitute present. For large projects it may be desirable to nominate substitutes at the start of the study.

Ideally meetings are not scheduled to last more than half a day at a time, as experience has shown that efficiency declines after long periods of intense mental activity. For the same reason, Hazop meetings should always take place in large well-ventilated rooms with provision for display of drawings and so on. At least one complete refreshment break should be arranged per session.

CARRY OUT THE STUDY

As part of the introduction to the study it is advantageous to go through the following preliminaries:
- reiterate the terms of reference, scope of work and programme;
- provide a brief revision of the Hazop technique for those who may be unfamiliar with the method;

- give a presentation describing the facilities to be studied, perhaps with the help of a model. Each P&ID should also start with a brief process review;
- agree the split of the P&ID into sections or lines for study.

Time spent at the start of the study on the above points will save a great amount of time during the study. It is particularly important that the chairman ensures that all team members are fully aware of the nature and purpose of the Hazop study, as it quite likely that some of them will not have been involved before. This could well require some effort in 'selling' the technique to the more sceptical members.

Figure 4.1 on page 52 shows a flowsheet for the application of Hazop to a typical continuous process operation. After the section to be studied has been selected, the chairman confirms critical parameters such as design flows, temperatures and pressures together with equipment and pipework specifications to ensure that everyone is aware of the design intent. This is often best done by means of a presentation by the process specialist.

It is the duty of the chairman to take the team through the guide words in a structured manner. The usual approach involves applying the parameter-related guide words to each section and then applying the general guide words to the plant as a whole or to much larger sections. By the nature of the guide words the technique has a considerable degree of self-checking built in. Thus, if a problem is missed under MORE FLOW there is a good chance that it may be picked up under, say, MORE PRESSURE or MORE TEMPERATURE.

Some form of check-list or prompt sheet is needed to ensure that nothing is missed; the list of guide words can be used for this purpose. Some organizations use pro-forma check-lists on which the chairman ticks off each guide word or deviation as it is completed. Computerized prompts are also available.

Final decisions on each and every action are taken by the whole team; the chairman does not have any prerogative. The study is a co-operative effort with the chairman acting as part of the team. For small studies the duties of chairman and secretary may be combined.

Once the study starts the chairman ensures that the guide words are followed in a rigorous and structured manner. It is very easy to get side-tracked as individual team members home in on one aspect of their discipline. The attitude of the team must at all times be positive and constructive. Hazop studies are extremely intensive and can take appreciable time. It is up to the chairman to maintain enthusiasm and motivation.

When carrying out a Hazop study it is not essential to seek solutions to all identified hazards. Hazop is primarily an identifying technique, not a problem-solving one. A great deal of time can be lost if the team attempts to find solutions to complex problems. It should simply identify areas which merit

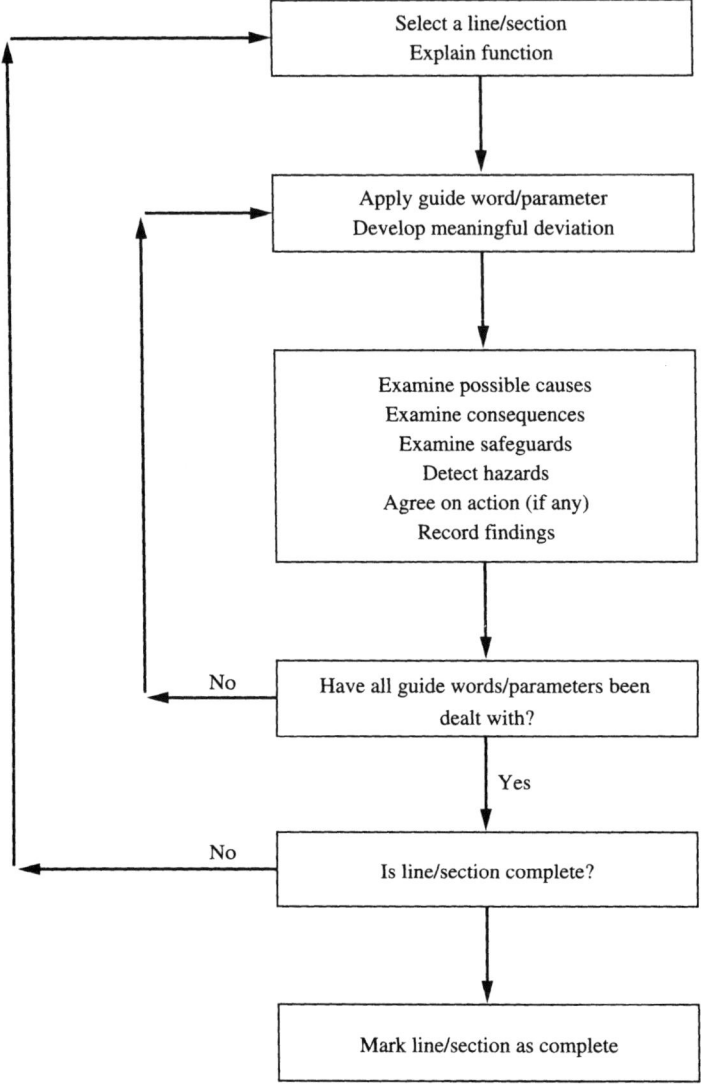

Figure 4.1   Hazop flowsheet.

further consideration and place actions and/or recommendations on others to come up with a solution. On the other hand some solutions may be self-evident and can form part of the firm recommendations from the study. Similarly, if it is not possible to obtain an answer to a question immediately, it should be referred outside the meeting as an action.

The 'actions' or 'recommendations' are an important part of any Hazop study. They take several forms including:

● requests for information on features of the design not known by the team — for example, 'will a particular relief valve handle a particular flow';

● note of the need for additional safety features to be engineered — for example, 'add a high pressure trip system';

● requests for a quantitative assessment to be carried out — for example, 'confirm that the failure rate of a protection system is acceptable';

● requirement of notes/warnings to be added to operating instructions.

All actions must be replied to and the reply considered by the chairman or some other experienced person. This matter is discussed below.

RECORD THE RESULTS

As the study progresses it is the responsibility of the secretary to record the discussion accurately. The chairman ensures that adequate time is allowed and helps by summarizing the discussion and the agreed actions/recommendations. The chairman ensures that all the team are in agreement with the findings recorded.

The basic record or work sheet is completed by the team secretary as the study proceeds using a document of the form shown in Figure 4.2 on pages 54–55. A number of formats are used, some omitting the safeguards column. Computerized systems are now available to produce reports interactively. The record sheets are completed as the study proceeds and actions/recommendations issued the next day. Provided that record sheets are clearly written, some organizations do not insist on typed sheets. The use of A3 size sheets can help to ensure that all the information is recorded in a clear and legible manner. The use of pro-forma action sheets with space for reply and reviewer's comments (Figure 4.3, page 56) can greatly expedite the process.

The Hazop record sheet forms part of the final plant documentation and should be in a form that can be submitted to the regulatory authorities. Reporting by exception — that is, only reporting those deviations which have serious consequences — is discouraged as it will not give a true and complete record of the study. It also makes auditing difficult and reduces the value of the study in any future accident investigation.

In order to keep track of the sections being studied, they are first identified on the master copy of the P&ID and, as each section is completed, is clearly marked as complete. In this way, it is possible to see by examining the drawing exactly what has been studied and the order in which the study has been carried out. Thus any branches, vents, drains and so on which may not have been studied can be clearly identified as such.

53

The final report includes the terms of reference, the scope of work, record sheets, confirmation that all actions have been cleared and a final conclusion by the chairman. The working copies of the P&IDs, the completed action reports and other drawings and documents used in the study are retained in a back-up file.

FOLLOW-UP

Following a Hazop design changes are inevitable, some of them perhaps as a result of the Hazop itself. At some appropriate stage in the project a further review should be carried out, preferably by the same chairman who conducted the original study. This review has three objectives:

• to ensure all changes made since the Hazop have not impaired the integrity of the original study;

• to review information, particularly vendor data, which may not have been available at the time of the original study;

HAZARD AND OPERABILITY STUDY REPORT

Project title:

Project number:

P&ID number:

Line number:

Guide word	Deviation	Cause	Consequences	

Figure 4.2   Hazop record sheet.

- to ensure that all actions/recommendations arising from the study have been cleared.

A large study may create several hundred actions and it is essential to set up some form of monitoring and control system to keep track of all the actions and the replies.

It is possible that some actions/recommendations cannot be resolved at this stage as they may still be awaiting vendor data or may concern documents not yet produced. In such cases a holding file is produced which is reviewed at the time of the pre-commissioning safety review.

Some actions may have to be resolved by other means, including quantitative hazard analysis, cost benefit analysis and application of the principles of ALARP. Although the Hazop team will not be directly involved in such studies, there are undoubtedly advantages in keeping the team informed.

A follow-up review report is generated indicating those issues that have been resolved and highlighting those outstanding.

Sheet of					
Date:					
Chairman:					
Study team:					
Safeguards	**Action**				
	Number	**By**	**Details**		**Reply accepted**

HAZOP ACTION SHEET		
Project:	**Project no:**	**Action no:**
P&ID no:	**Date:**	**Tape ref:**
Action on:	**Date for reply:**	
Description:		
Reply: Signed: Date:		
Review comments: Accepted/rejected (Leader) Date:		
Issued	Returned	Complete
Return completed form to:		

Figure 4.3   Hazop action sheet.

4.4   COMPUTERIZED REPORTING SYSTEMS

The task of the Hazop secretary is onerous and the amount of paperwork generated by a study is considerable. Thus Hazop is an obvious candidate for computerization. A number of systems have been developed both within companies and by consultants. They are mostly based on word processing packages and generally include prompts to assist the chairman in the conduct of the study. They can produce the equivalent of the manual Hazop record or work sheet and can generate the actions/recommendation requests. In addition to the recording and reporting functions, the use of a computer-based system can have the following advantages:

- automated recall of parameters and sequential review of guide words;
- a tools package to review commonly occurring fault conditions;
- a projection system, so that all team members can see what is being recorded at the time;
- a simple method of ranking and prioritizing actions;
- comprehensive Hazop tutorial.

Most systems mirror as closely as possible the accepted methodology of the Hazop study. Although a set of guide words, parameters and deviations are included with most packages, it is usually possible to customize them before the start of the study. Thus the chairman can delete guide words not applicable and add any specific to the process being studied.

Virtually all systems are PC-based; they do not require excessive computing power but they need reasonable memory and a hard disk drive. They can all produce files for further editing. In order to make maximum use of computer-based systems it is desirable that the whole team can see the screen. This can be done using one of the commonly available projection systems.

The output varies from package to package but most can produce a series of reports and generate action sheets sorted in any required manner. All commercial packages also have the facility to control actions; thus the responses to the actions can be input to the system as they are received. A notebook facility is usually included, offering some word processing capability. This can be used to record useful information not covered by the structured guide words.

A typical set of reports would be:

- whole report of study;
- report by date;
- brief project response date;
- full report (including responses);
- severity/likelihood report;
- priority report;
- notebook report.

The computer is intended to be a tool used to support the study. It helps the team members focus on the actual study and not become tied up with unnecessary system complexities. The use of computer-based systems can do much to reduce the secretarial and administrative work, assist the chairman by providing prompts and ensure control of actions/recommendations. Several attempts have been made to produce 'expert systems' for Hazop but none so far have been able to replace the traditional Hazop team.

4.5 HAZOP OF BATCH PROCESSES

The original Hazop technique related to continuous processes wherein, between start-up and shutdown, there are long periods of steady state operation. For batch processes the situation is much more complex because of the time variable. The status of the plant is constantly changing in some established sequence. Since the P&ID does not sufficiently define the system, a set of operating instructions and some form of sequence chart are also needed. Because plant operators generally play a larger part in batch operations, more consideration has to be given to human operator reliability. Similarly the increasing use of computers in the sequencing of batch operations requires more attention to be given to the whole question of software safety assurance.

In order to carry out a Hazop study of a batch process, not only is more documentation needed and more preparation required, but the technique itself has to be modified somewhat. Firstly, some of the guide words have to be viewed in a different light. The parameter FLOW should be considered in terms of 'quantity' rather than, or as well as, 'rate'. An additional parameter LEVEL is often needed and it may be necessary to consider QUANTITY as an independent parameter.

It is also necessary to repeat the review several times for the same plant item. Thus, a reactor may first be studied during the addition of the reactants, repeated whilst it is brought up to reaction temperature and again for the reaction stage itself when, say, an initiator is added. It is essential to determine the status of the plant between successive operations. It is also important to realize that operations may be taking place in different parts of the plant at the same time. Whilst the reaction is in progress a second batch of reactants may be prepared in a premix tank. The Hazop study must consider the possibility of an interaction between the two steps.

So to carry out a batch Hazop, the chairman first ensures that all necessary information on sequencing is available, then divides the process up into a clearly defined set of operations. The guide words and parameters are customized for the process concerned. The Hazop P&IDs are clearly marked up showing the steps or sequences to be studied and each step then checked off on completion. More details of the application of the technique to batch processes are given in Reference 2.

4.6 EXTENSIONS OF HAZOP

The basic Hazop technique can be extended to be used in non-process situations and to process-related applications at different stages of development. This requires the formulation of a special set of guide words. To apply the technique to the effect of process operations on, say, nearby office accommodation, the following guide words could be used:

- smoke hazard;
- toxic gas hazard;
- flammable gas hazard;
- missiles;
- explosions;
- fire;
- noise;
- smell.

The technique has also been successfully applied to electrical and mechanical systems, again with suitable modifications to the guide words. Examples are given in Reference 4.

4.7   HAZOP DEMONSTRATED

To demonstrate Hazop, consider a simple hydrocarbon feedstock transfer example as shown in Figure 4.4 on pages 60–61. This example is taken from Reference 1 with the kind permission of the author, Dr H.G. Lawley.

The hydrocarbon is transferred from intermediate storage via the J1 pumps and a 1 km overground pipeline running adjacent to a public road into a 25 m^3 nitrogen blanketed feed/settling tank operating at 20°C and 1 bar g. Control is on liquid level and there is a split range pressure control from the 2 bar g site nitrogen supply. Integrated flow measurement is provided at J1 pump common delivery for accountancy purposes and a manual sample point is provided at the tank inlet for off-line analysis. There is also a branch, normally closed, to a petrol blending system.

Figure 4.5 on pages 62–65 gives an action report which would be expected of a typical team. Note that there is no 'safeguards' column in this report and a total of 20 actions were generated. This is a typical number for a single plant section.

4.8   CONCLUSIONS

The Hazop technique has now been proven as a method of safety assurance in the process and related industries and is accepted throughout the world as a way of demonstrating that a project has been subjected to a rigorous safety examination. It is generally agreed[2] that success or failure depends upon:

- the accuracy of drawings and other documents used in the study;
- the expertise and experience of the team;

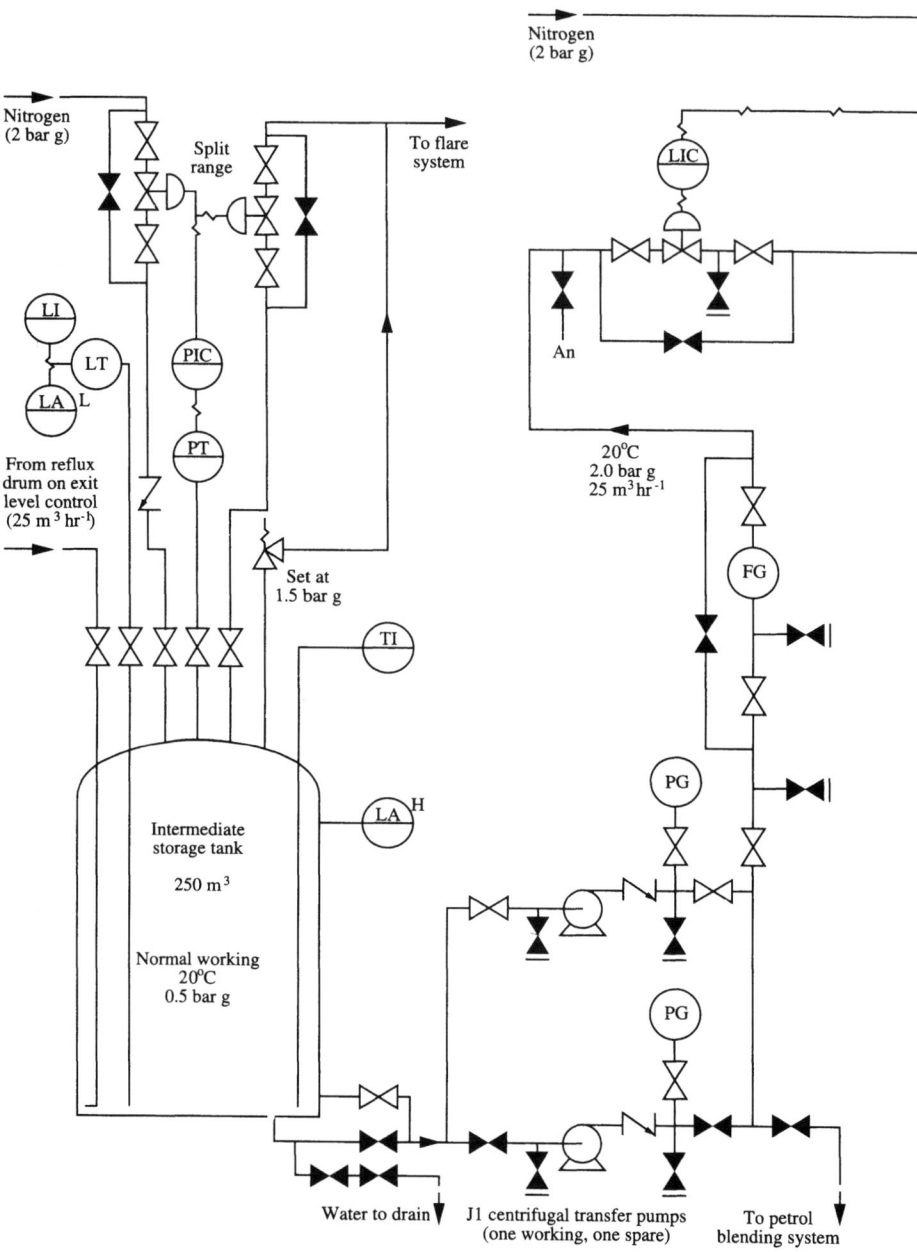

Figure 4.4   Hydrocarbon transfer system.

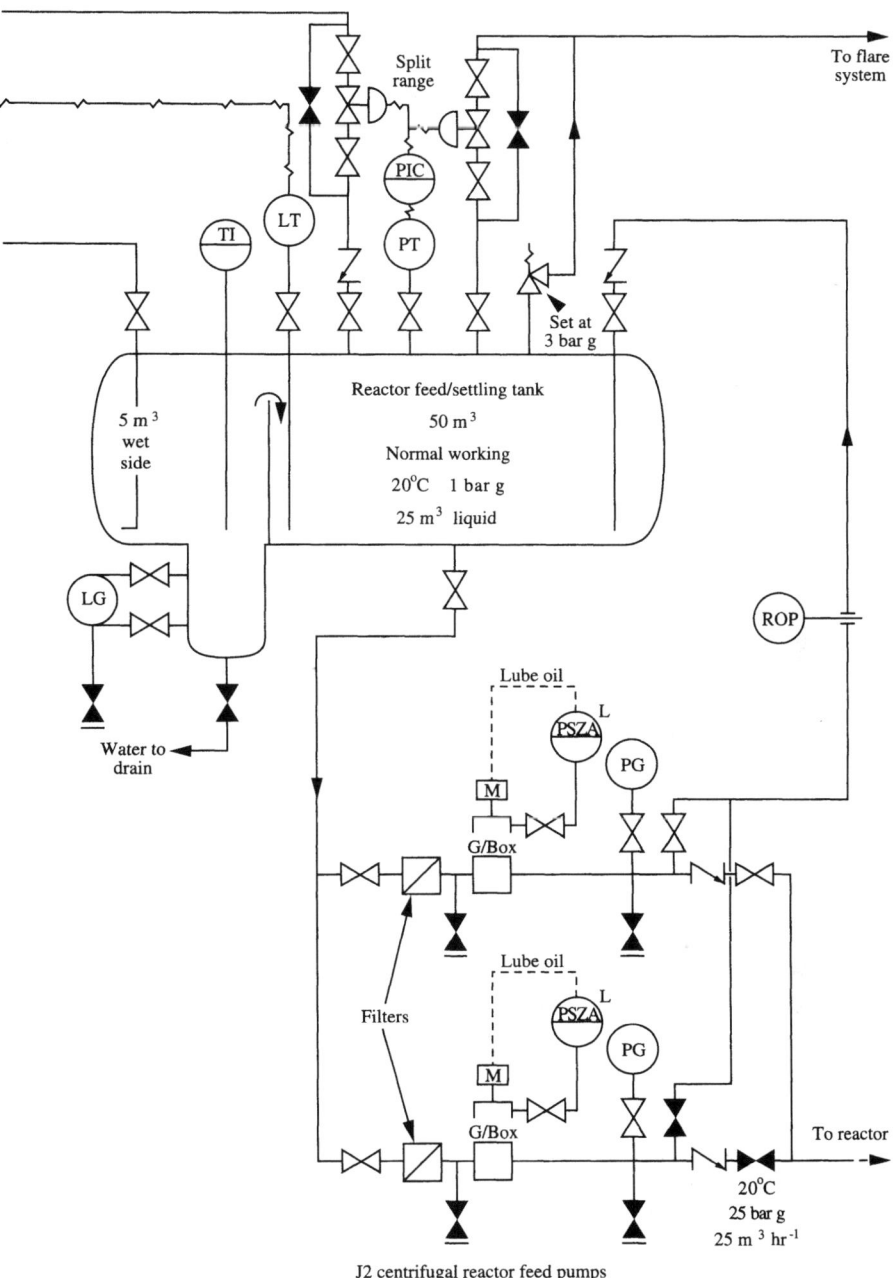

Split range

To flare system

PIC

TI   LT

PT

Set at 3 bar g

Reactor feed/settling tank
50 m³
Normal working
20°C   1 bar g
25 m³ liquid

5 m³
wet
side

LG

ROP

Water to drain

Lube oil

L
PSZA

PG

M

G/Box

Filters

Lube oil

L
PSZA

PG

M

G/Box

To reactor

20°C
25 bar g
25 m³ hr⁻¹

J2 centrifugal reactor feed pumps
(one working, one spare)

HAZARD AND OPERABILITY STUDY OF PROPOSED HYDROCARBON TRANSFER SYSTEM:
Results of line section from intermediate storage to buffer/settling tank

Guide word	Deviation	Possible causes	
NONE	NO FLOW	(1) No hydrocarbon available at intermediate storage	
		(2) J1 pump fails (motor fault, loss of drive, impeller corroded, etc), power failure	
		(3) Line blockage, LCV fails shut	
		(4) Line fracture	
		(5) Valve closure in error	
REVERSE	REVERSE FLOW	(6) Failure of PIC and higher-than-normal N_2 pressure	
		(7) Failure or leakage of NRV	
		(8) Backflow through standby pump of standby pump	
MORE OF	MORE FLOW	(9) LCV fails open or LCV bypass open in error or both pumps operating	

Figure 4.5(a) Specimen Hazop report, Page 1.

Consequences	Action required
Loss of feed to reaction section and reduced output. Polymer formed in heat exchanger under no flow conditions. Buffer tank level falls.	(a) Ensure good communications with intermediate storage operator (b) Install low level alarm on settling tank LIC
As for (1)	Covered by (b) Add pump running indicator lights (also in control room)
As for (1) J1 pump overheats Rising level in storage tank?	Covered by (b) (c) Install kickback on J1 pumps
As for (1) Hydrocarbon discharged into area adjacent to public highway	Covered by (b). Consider adding second FQ at buffer tank inlet. (d) Institute regular patrolling and inspection of transfer line.
As for (1)	(e) Review operator reliability and provision for J2 pump protection
N₂ gas breakthrough	(f) Add anti-siphon provision in buffer tank dip-pipe
N₂ passed to intermediate storage	(g) Check capacity of intermediate storage relief system. Review maintenance programme.
Reduced delivery to buffer tank	(h) Check operating instructions for isolation
Settling tank overfills	(j) Install high level alarm on LIC and check sizing of relief opposite liquid overfilling. Consider high-high level trip with auto re-set. (k) Institute locking-off procedure for LCV bypass when not in use
Incomplete separation of water phase in tank leading to reaction problems	(l) Extend J2 pump suction line to 300 mm above tank base

Guide word	Deviation	Possible causes	
LESS OF	LESS FLOW	(10) Leaking flange or valve stub not blanked and leaking	
		(11) Blocked pump section	
		(12) Impeller wear	
		(13) Possible vapour locking in hot weather?	
MORE OF	MORE PRESSURE	(14) Isolation valve closed in error or LCV closes, with J1 pump running	
		(15) Thermal expansion in an isolated valved section due to fire or strong sunlight	
MORE OF	MORE TEMPERATURE	(16) High intermediate storage temperature	
LESS OF	LESS TEMPERATURE	(17) Winter conditions	
PART OF	HIGH WATER CONTENT IN STREAM	(18) High water level in intermediate storage tanks	
	PRESENCE OF MORE VOLATILE COMPONENTS	(19) Disturbance on distillation columns upstream of intermediate storage	
OTHER	MAINTENANCE	(20) Equipment failure, flange leak, etc	

Figure 4.5(b) Specimen Hazop report, Page 2.

64

Consequences	Action required
Material loss adjacent to public highway	Covered by (d)
Pump damage	(m) Check filter periodically
Reduced delivery pressure	(n) Review inspection/maintenance programme
As for (15)	(o) Provide suitable vent on pump casing
Transfer line subjected to full pump delivery, surge pressure or closed head pressure. Pressure exceeds line specification. Slow closing time of control valve.	(p) Covered by (c) except when kickback blocked or isolated. Check line, FQ and flange ratings, and reduce stroking speed of LCV if necessary. Install a PG upstream of LCV and an independent PG on settling tank.
Line fracture or flange leak	(q) Install thermal expansion relief on valved section (relief discharge route to be decided later in study)
Higher pressure in transfer line and settling tank	(r) Check whether there is adequate warning of high temperature at intermediate storage. If not, install a device.
Water sump and drain line freeze-up	(s) Lag water sump down to drain valve, and steam trace drain valve and drain line downstream
Water sump fills up more quickly. Increased chance of water phase passing to reaction section.	(t) Arrange for frequent draining off of water from intermediate storage tank. Install high interface level alarm on sump.
Higher system pressure	(u) Check that the design of settling tank and associated pipework, including relief valve sizing, will cope with sudden ingress of more volatile hydrocarbons
Line cannot be completely drained or purged	(v) Install low-point drain and N_2 purge point downstream of LCV. Also N_2 vent on settling tank.

- the ability of the team to visualize deviations, causes and consequences;
- the ability of the team to assess the seriousness of hazards;
- the skill of the chairman in keeping the study on track.

Of these the skill of the chairman is perhaps the most important. The technique will only be successful if the chairman has the necessary training and experience for the task. However, even the most skilled and experienced chairman will not be able to compensate for a badly constituted team.

REFERENCES IN CHAPTER 4
1. Lawley, H.G., 1974, Operability studies and hazard analysis, *Chem Eng Prog*, 70 (4): 45–56.
2. Chemical Industries Association, 1977, *A Guide to Hazard and Operability Studies*.
3. Turney, R.D., 1990, Designing plants for 1990 and beyond: Procedures for the control of safety, health and environmental hazards in the design of chemical plant, *Trans IChemE*, 68 (B1): 12–16.
4. Bullock, C.J., Mitchell, F.R. and Skelton, R.L., 1990, *Operability Study Applications in Selected Work Field Areas. Safety and Reliability in the 90s* (Elsevier, New York, USA and London, UK).

FURTHER READING
1. Kletz, T., 1992, *Hazop and Hazan: Identifying and Assessing Process Industry Hazards*, 3rd edition (IChemE, Rugby, UK).

5. FAILURE MODE AND EFFECT ANALYSIS (FMEA)

Failure Mode and Effect Analysis (FMEA) is an alternative method of hazard identification: it is less formalized than Hazop and involves the consideration of the possible outcomes from all discerned failure modes or deviations within a system. It is particularly suited to complex mechanical or electrical systems and can be applied at different levels of detail or complexity, which are usually referred to as the 'hierarchic' or 'indenture' level.

FMEA systematically identifies the consequences of component failure on that system and determines the significance of each failure mode with regard to the system's performance. The technique is primarily used to study material and equipment failure and can be applied to a wide range of technologies. It is a 'bottom up' technique wherein each failure mode within the system is traced forward logically in sequence to the final effect.

The use of FMEA is generally limited by the time and resources available and also by the availability of the necessary data. It is thus best confined to items shown to be critical by earlier analyses or by other criteria such as safety, cost, accessibility and so on. It is often used in conjunction with Fault Tree Analysis (FTA). In order to ensure the appropriate systems are analysed, it is advisable to start at the highest possible (or simplest) 'hierarchic' or 'indenture' level of a system. A broad, probably qualitative, analysis at this level identifies the most important or critical contributors. Once identified such areas can then be analysed at the next or more detailed hierarchic level, at which stage it should be possible to become more quantitative in the approach.

5.1   METHODOLOGY OF FMEA

The following steps are required:
- define the system to be evaluated, the functional relationships of the parts of the system and their performance requirements;
- establish the level of analysis;
- identify failure modes, their cause and effects, their relative importance and their sequence;
- identify failure detection, rectification and isolation provisions and methods;
- identify design and operating provisions against such failures;
- summarize, recommend corrective actions and issue report.

DEFINITION OF SYSTEM TO BE EVALUATED

As with all safety analysis, it is first necessary to define the extent of the system to be analysed. Because of the complexity of the technique, FMEA is usually performed in relatively small steps. It can only be carried out by analysts with a knowledge of the system concerned and the ways in which its components interact.

LEVEL OF ANALYSIS

The choice of level of analysis can be very difficult. One method is to perform an FMEA based on the functional structure of a system rather than on its physical components. In a functional FMEA the failure modes are expressed as failure to perform a particular subsystem function: it should consider both primary and secondary functions. The primary function is that for which the subsystem was provided, whereas the secondary function is one which is merely a consequence of the subsystem's presence. For a comprehensive discussion of this topic, see BS 5760 Pt 5, 1991[1].

ANALYSIS OF FAILURES

All possible failure modes should be considered. Examples include:
- premature operation;
- failure to operate when required;
- intermittent operation;
- failure to cease operation when required;
- loss of output or failure during operation;
- degraded output.

Having identified the various failure modes, the analyst then looks at the likely causes and the effects on both the component concerned and on the system of which it is a part. In doing this, consideration is given to the relative importance of the effects and the sequence in which they occur. The safeguards against such failures and methods of detecting them are then examined.

REPORTING

With the above information, it should be possible to identify the most significant failures in terms of their effects on the overall system, and to decide whether or not the existing safeguards and detection devices are adequate. Having identified a weak link it may be necessary to subject that component to more detailed analysis, or perhaps it may be decided to eradicate or reduce the probability of failure by redesign. There is no standard format for FMEA reports but examples of typical reports are given later in Tables 5.1 and 5.2 on pages 72 and 76–77.

5.2 CRITICALITY ANALYSIS

The technique can be extended to identify semi-quantitatively those areas which are critical from a safety or reliability point of view. Criticality is defined in the same way as risk — that is, a combination of the severity of an effect and the probability or expected frequency[1]. The simplest approach requires a form of ranking or quantification in terms of effect and frequency. Effects are normally ranked into one of the following categories:
- loss of mission due to inability of equipment to perform;
- economic loss due to lack of output or function;
- damage to plant or third party property;
- injury to operating personnel or the public;
- death to operating personnel or the public and/or significant damage to the environment.

An alternative ranking[2], in reverse order of severity is:
- catastrophic — may cause death or total system loss;
- critical — may cause severe injury or damage;
- major — may cause some injury or damage;
- minor — requires unscheduled maintenance.

Clearly all such classifications are subjective and depend very much on the system under consideration.

The quantification of frequency depends on the data available and may again be a simple ranking, such as one depending on failure probability during the operating time interval:
- extremely unlikely — 0.001;
- remote — between 0.001 and 0.01;
- occasional — between 0.01 and 0.1;
- reasonably frequent — between 0.1 and 0.2;
- frequent — 0.2.

If more precise failure rate data is available — for example, actual failure rates — it should be used.

A ranking system as employed in Hazop can be used instead. Care is required in the use of any ranking system, to ensure that the order of ranking is consistent: the highest numerical ranking may be given to the most desirable outcome rather than to the least desirable. This reverse order (highest least desirable) is frequently found in FMEA.

The effect (or consequence) and frequency rankings are then added to the report form and a criticality matrix is produced from which the most critical component can be identified. In this way limited resources can be applied to the greatest advantage.

CALCULATION OF FAILURE MODE CRITICALITY

A rather more mathematical criticality analysis can be used when actual failure rate data are available. The sources and reliability of such data are discussed in detail in Chapter 7. For each entry the predicted item failure rate, λ_p, is multiplied by a failure rate apportionment factor, α, to obtain a failure rate for a particular mode, λ_m. This is then multiplied by a failure effect probability factor, β, which represents a subjective engineering judgement of the likelihood of the failure mode causing the undesired effect. The result is a failure mode frequency, λ_f, which again may be used to identify critical components and so indicate priorities for corrective action. The value of λ_p may also be corrected for any environmental factors (see Chapter 7).

The value of β may be quantified as given below:

Failure effect	β
Actual loss	1.00
Probable loss	0.1 to 1.00
Possible loss	0 to 0.1
No loss	0

Using the above notation the value of λ_f will be in the same units as the failure rate λ_p. In some cases the failure rate per mission or number of operating cycles is required, in which case the equation is further multiplied by a time or cycle factor, t.

FMECA is less rigorous than Quantitative Risk Analysis (QRA) in that the effects are expressed in terms of severity level rather than being quantified in detail. The results of FMECA are often expressed in the form of criticality bands[1] as shown in Figure 5.1. On this basis band A is acceptable, bands B and C are increasingly unacceptable and band D could well be totally unacceptable.

5.3   CORRECTIVE ACTION AND FOLLOW-UP

Based on the FMEA, the following courses of action are normally open:
- eliminate cause of failure;
- reduce probability that the cause of failure will result in the failure mode;
- reduce severity of failure by redesign or add protection redundancy;
- increase probability of detection;

Figure 5.1   FMECA criticality matrix.

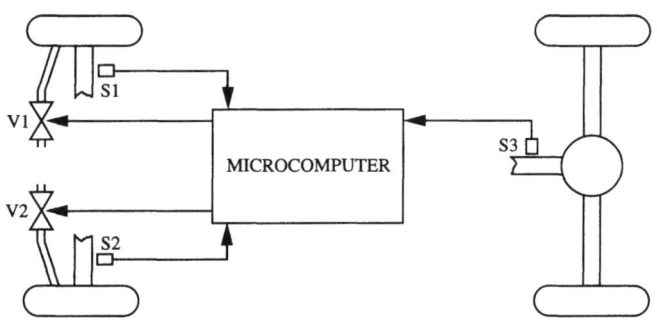

Figure 5.2   Passenger car anti-skid braking system.

5.4   EXAMPLES.

ANTI-SKID BRAKING SYSTEM (FUNCTIONAL FMEA)
Figure 5.2 shows a simple anti-skid braking system (ABS) for a rear wheel drive passenger car. The objective is to prevent locking of the front wheels during heavy braking under bad road conditions.

Speed sensors S1 and S2 measure the speed of the two front wheels and S3 measures the speed of the drive shaft which gives a measure of the speed

71

TABLE 5.1

Passenger car anti-skid braking system FMEA

Component	Failure mode	Failure effect(s)	Comment
Front wheel sensor, S1 or S2	No output signal	Micro-computer will assume one wheel has stopped, and send a signal to open the relief valve on that wheel. Partial loss of front wheel braking.	Uneven braking on front wheels. Alarm system required — switch off micro-computer?
Front wheel valve, V1 or V2	Fail to open	One front wheel could lock on heavy braking	Unrevealed failure mode. Test facility required?
	Fail to close	Partial loss of front wheel braking	Uneven braking on front wheels. Isolation/stop valve required?
Rear wheel sensor, S3	No output signal	Micro-computer will have no reference speed from rear wheels, and hence will not attempt to open V1 and/or V2. Both front wheels could lock on heavy braking.	Alarm system required
Micro-computer, MC	No output signals to either front wheel valve	Both front wheels could lock on heavy braking	Alarm system required?
	No output signal to one front wheel valve	One front wheel could lock on heavy braking	Alarm system required?
	Spurious output to both front wheel valves	Total loss of front wheel braking	Alarm system required — switch off micro-computer?
	Spurious output to one front wheel valve	Partial loss of front wheel braking	Alarm system required — switch off micro-computer?

of the rear wheels. Signals from the three speed sensors are fed to a micro-computer. If the speed of one of the front wheels falls significantly below the speed of the rear wheels, indicating locking, then the appropriate valve V1 or V2 is opened to reduce braking force until adhesion is regained.

The FMEA takes the six major components — sensors S1, S2, S3, valves V1 and V2 and the micro-computer — and considers their respective failure modes and the effects. The sensors are probably most likely to fail so that they give no output, the design being such that high or low output is unlikely although this could be considered later. The valves could fail to open on demand or fail to shut when required. Total brake fluid leakage is a separate matter outside the scope of this analysis which is limited to the ABS. The micro-computer and associated electronics could fail to give the required output signal or give a signal when not required.

The results of a simple FMEA are shown in Table 5.1. A comment column is provided to list possible modifications but these in turn should be examined to make sure that they do not make matters worse. As with most other forms of safety analysis, multiple failures (double jeopardy) are not considered.

PILOT-OPERATED RELIEF VALVE (COMPONENT FMEA)

Figure 5.3 on page 74 (taken from Reference 2) shows a pilot-operated relief valve common in many high pressure installations. The main valve is operated by a small pilot valve mounted above the main valve body and connected to it by external supply and exhaust tubes. This pilot valve controls the pressure on the piston which holds the main valve on its seat. This type of valve reduces the load on the spring needed to hold the valve closed and should result in more rapid re-seating after use. It does, however, contain many more components than a simple spring-loaded relief valve and hence can be less reliable. The object of the FMEA in this case is to identify any weak links in the valve assembly.

Table 5.2 on pages 76–77 shows the results. The format is slightly different from the previous example although the basic purpose is the same. The report shows that the pilot valve and the dipper tube are the critical items and that care needs to be taken in the design of the pilot valve. This example shows the detail to which component FMEA can be taken, but analysis at this level is only really justified for safety critical components.

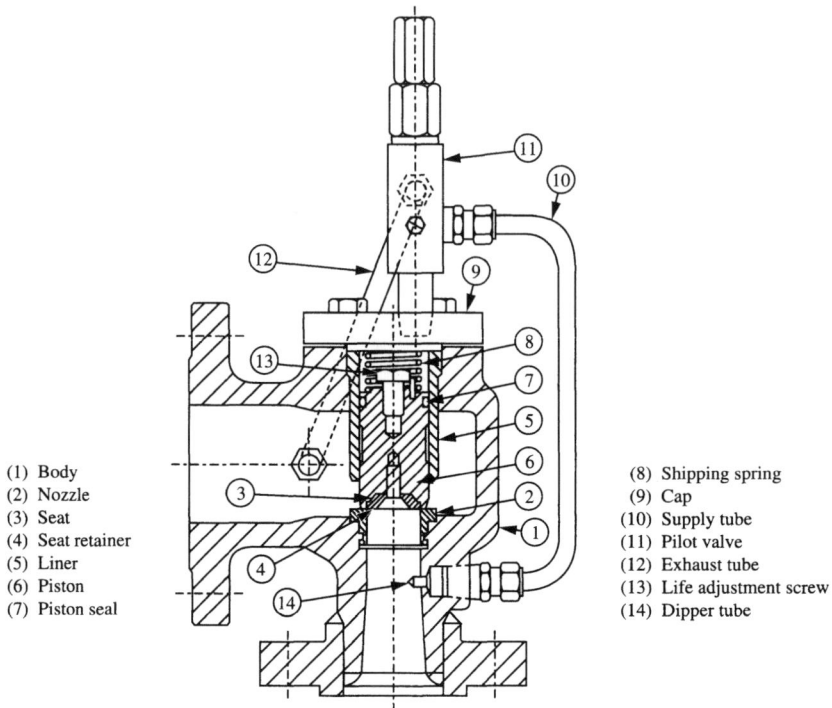

(1) Body
(2) Nozzle
(3) Seat
(4) Seat retainer
(5) Liner
(6) Piston
(7) Piston seal

(8) Shipping spring
(9) Cap
(10) Supply tube
(11) Pilot valve
(12) Exhaust tube
(13) Life adjustment screw
(14) Dipper tube

Figure 5.3   Pilot-operated relief valve.
Reproduced with permission of Mechanical Engineering Publications, from *Reliability of Mechanical Systems*, 2nd edition, 1994.

SEMI-QUANTIFICATION OF ABS EXAMPLE

The application of criticality analysis can be shown using the anti-skid braking system (ABS) previously discussed. The effects could be ranked for severity as follows:

Total loss of front wheel braking	5
Uneven front wheel braking	4
Both front wheels lock on hard braking	3
One front wheel locks on hard braking	2

This ranking may be debatable and illustrates one of the problems with the technique — it relies very much on the experience and judgement of the analyst. It is assumed that the analyst in this case considered that total locking of one or both wheels would simply leave the car in the same condition as one without ABS and that a skilful driver should be able to compensate. But even a skilful driver could do little if the valves remained open. The severity ranking and frequency data are shown in Table 5.3 (pages 78–79) and Table 5.4 (page 78) gives the ranking matrix. It is clearly seen from this analysis that the weak links in the system are the front wheel valves V1 and V2 (criticality band C). Thus attention should be paid to ensuring the reliability of these components and perhaps providing some testing system to give the driver confidence that they will work when called upon.

FURTHER EXAMPLE OF CRITICALITY ANALYSIS

Figure 5.4 on page 80 shows a forced lubrication system for a process compressor. The system comprises a tank with a heater to ensure that the oil is fluid at low ambient temperatures and two pumps, one jet and one low speed. The high pressure pump is protected by a pressure relief valve. Table 5.5 on page 80 shows a complete FMECA for the system. The value of λ_p is taken from the literature but the values of α and β are based on the judgement of the analyst. Thus if the sump heater goes 'open circuit' there is a probability of 1 that it will not heat the oil, but the probability of loss of the compressor will depend on other circumstances such as the ambient temperature and the oil temperature at the time of failure. The analyst has thus assigned a β factor of 0.1 in this case. The relief valve sticking open will reduce the available flow but probably not enough to cause failure. A blockage of either pump inlet may be partial and not stop the total flow — hence the β factor of 0.8 — but a loss of lubrication would lead to compressor failure resulting in a β value of 1. The analysis shows that the low speed pump is the weak link in the system.

5.5   CONCLUSIONS

FMEA allows a flexibility of approach — functional or hardware based — and the depth of study can be limited to suit the objectives. It enables rapid identification of critical areas allowing effort to be concentrated where it will achieve the greatest results. It can be combined with other techniques such as Fault Tree Analysis. It is most suited to complex mechanical or electrical systems but it does have applications in the process industries, particularly as instrumentation and control systems become more complex and increasing reliance is placed upon such systems for safe operation.

TABLE 5.2
Pilot-operated relief valve FMEA[†]

Item/component	Failure mode	Local failure effects
Seat	Scored or damaged	Relief valve leaks
Piston seal	Leaking or damaged (eg, chemical reaction)	Pressure above piston drops and relief valve opens spuriously
Piston liner	Scored or damaged	As above
Shipping spring	Weakened, worn or fatigued	Reduction in downwards force on piston may cause spurious, intermittent opening of relief valve
Supply tube	Leakage at connections	Reduction in supply pressure to pilot valve will cause increase in pressure required to activate relief valve
Pilot valve	(i) Leakage	Greater pressure required to activate relief valve
	(ii) Valve seized shut	Pilot valve unable to activate opening of relief valve
Exhaust tube	Leakage at connections	After relief valve has been opened, unable to establish pressure equilibrium required to reclose valve
Dipper tube	Blockage	Loss of supply pressure to pilot valve. Pilot valve unable to activate opening of relief valve.

[†] Reproduced with permission of Mechanical Engineering Publications, from *Reliability of Mechanical Systems*, 2nd edition, 1994.

System failure effects	Symptoms	Rectification	Corrective action
Hydrocarbon gas leaks from system and pressure falls	Noise of leaking gas. Pressure reduction.	Replace valve	
Hydrocarbon gas spuriously vented from system	Spurious opening of relief valve	Replace valve	Choose materials resistive to chemical attacks, etc
As above	As above	As above	Abrasion-resistive materials for piston lining
As above	As above	As above	Choose material with suitable fatigue characteristics
Hydrocarbon gas allowed to rise to a higher pressure than specified. May be critical.	Gauges indicate unrelieved pressures	Replace valve	Ensure connections are clean and adequately sealed
As above	As above	As above	As all of those above
Hydrocarbon gas pressure allowed to rise to dangerous levels. Will be critical.	Gauges indicate dangerously high pressures	As above	Careful design of pilot valve should reduce occurrence of this event to a minimum
Once opened, relief valve permanently open and hydrocarbon gas continually vented from system	Relief valve fails to reclose	Replace valve	Ensure connections are clean and adequately sealed
Hydrocarbon gas pressure allowed to rise to dangerous levels. Will be critical.	Gauges indicate dangerously high pressure	As above	Some form of filtering may be considered to prevent blockage due to foreign objects

TABLE 5.3
Anti-skid braking system FMEA (semi-quantified)

Component	Failure mode
Front wheel sensor, S1 or S2	No output signal
Front wheel valve, V1 or V2	Fail to open
	Fail to close
Rear wheel sensor, S3	No output signal
Micro-computer, MC	No output signal to either valve
	No output signal to one valve
	Spurious output to both valves
	Spurious output to one valve

TABLE 5.4
ABS FMEA frequency/severity matrix

Frequency	Severity				
	1	2	3	4	5
21 to 25 (5)		V1 V2 Fail to open			
16 to 20 (4)				V1 V2 Fail to close	
11 to 15 (3)		MC No output to one V			
6 to 10 (2)			S3 No output	S1 S2 No output	
1 to 5 (1)			MC No output to either	MC Spurious output to one	MC Spurious output to both

Failure effect	Severity	Frequency*	Ranking
Partial loss of front wheel braking (uneven)	4	8	8
One front wheel lock on heavy braking	2	22	10
Partial loss of front wheel braking (uneven)	4	16	16
Both front wheels lock on heavy braking	3	8	6
Both front wheels lock on heavy braking	3	5	3
One front wheel locks on heavy braking	2	12	6
Total loss of front wheel braking	5	1	5
Partial loss of front wheel braking (uneven)	4	4	4

* Frequency data is the number of failures per 10^6 operating hours

REFERENCES IN CHAPTER 5

1. BS 5760, Part 5, 1991, Reliability of systems, equipment and components (BSI, London, UK).
2. Davidson, J. (ed), 1994, *Reliability of Mechanical Systems*, 2nd edition, Chapter 14 (IMechE, London, UK).
3. King, C.F. and Rudd, D.F., 1972, Design and maintenance of economically failure tolerant processes, *AIChE J*, 18 (2): 257.

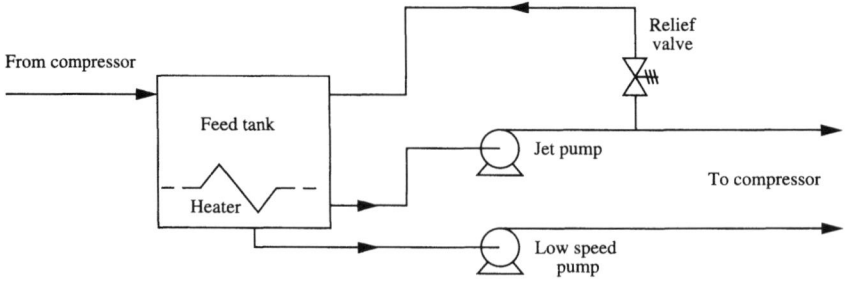

Figure 5.4   Compressor lubrication system.

TABLE 5.5
Effect: Loss of lubrication. Severity category: 3.

Component	Failure mode	Cause	λ_p	α	λ_m	β	λ_f
Heater	Fail to heat	Open circuit	2.0	1.0	2.0	0.1	0.2
Jet pump	RV open	Spring stuck	1.0	0.2	0.2	0.5	0.1
Jet pump	Low flow	Blockage	1.0	0.2	0.2	0.8	0.16
Jet pump	No flow	Drive failure	1.0	0.1	0.1	1.0	0.1
LS pump	Low flow	Blockage	5.0	0.2	1.0	0.8	0.8
LS pump	No flow	Drive failure	5.0	0.1	0.5	1.0	0.5

The failure rates are numbers of faults per 10^6 hours

6. BASIC QUANTITATIVE RISK ASSESSMENT (QRA)

Whilst qualitative and semi-quantitative assessments are very valuable in identifying hazards, they may not always give sufficient information to allow decision-making on complex and potentially hazardous processes. Under these circumstances it is necessary to carry out a full Quantitative Risk Assessment (QRA).

It will be recalled that risk can be expressed as a function of the frequency or probability of an incident and the consequences of that incident. QRA looks at both aspects and should produce sufficient information for comparison against acceptance criteria, and hence allow decisions to be made on a sound basis.

The risk analyst is concerned with the frequency/probability and with the consequence aspects and can advise on acceptance criteria, but in the end the decision about what is and is not acceptable is essentially a policy matter. The application of the ALARP principle and the questions of benefit — perceived or otherwise — must also be considered.

6.1 THE LOGIC TREE APPROACH

All QRA is based on the construction of logic diagrams — usually called trees — which show how the various causes of an incident are related. Failure data is then used to quantify the logic trees in order to arrive at a probability or frequency of that event. Two types of logic trees are in common use — fault trees and event trees. Fault trees are essentially top down logic trees in which the various events leading up to an incident are related through a series of logic gates. Event trees often follow on from fault trees and examine the events which result from an incident in terms of simple logic gates. A fault tree thus gives the probability of an untoward incident happening and an event tree examines the possible consequences of that incident.

6.2 PRINCIPLES OF QRA

In order to understand the basic principles of QRA it is first essential to clarify certain definitions:

Probability is a dimensionless number between 1 and 0, representing the probability of a particular state existing or the likelihood that one event will succeed another. It may therefore be time-related. The usual symbol is P or $P(t)$ if time-related.

Frequency is the number of times a particular event occurs per unit time (for example, the number of times a tank overflows per year). It has the dimensions of $(\text{time})^{-1}$ and is also referred to as the *failure or hazard rate*. Symbols used include F and λ.

Duration is the time during which a particular state exists (for example, the time during which the tank is actually overflowing). It has the dimensions of (time) and the units must be consistent with frequency. Symbols include D, T and τ.

Protective system is a system or device which is installed to stop the hazard occurring (for example, a high level trip for the tank).

Demand rate is the rate at which a protective system is required to take action (for example, the frequency of the level reaching the high level trip).

Failure is when a protective system is incapable of carrying out its duty. Systems can *fail* either to a *dangerous* condition (for example, the trip fails to stop the flow) or to a *safe* condition (for example, the flow is stopped before the high level is reached). There are two types of failure:
- *revealed*, in which faults show themselves by the effect they have on the system. They are self-evident and lead to immediate action;
- *unrevealed*, in which faults remain undetected until the affected equipment (for example, a trip system) is called upon to operate or is routinely proof tested.

Proof testing is a method of checking at regular intervals to ensure that the protective system is in full working order.

Fractional dead time is the fraction of the total time that the protective device is in a failed state and is a measure of the probability of its failure to provide the design intended protection. Usually simply referred to as *fdt* or sometimes as ϕ. Note that a more exact definition of fractional dead time is 'mean unavailability'.

6.3   FAULT TREE ANALYSIS

Fault Tree Analysis (FTA) is the most common and generally the most useful technique in Quantitative Risk Assessment. In FTA the undesired or 'top' event is first identified and the sequence of events which leads up to it is then examined and placed in a logical order. The events are linked by 'AND' or 'OR' gates depending on their relationship. FTA is only designed to deal with the situation in which a component or subsystem either works or fails; it does not deal in a satisfactory manner with the situation of degraded performance.

The OR gate describes a situation where the next event will occur if one or more of the input events exist (Figure 6.1). Hence any or all of the three input events shown in Figure 6.1 will give rise to the output event.

The AND gate describes a situation where the output requires the simultaneous existence of all the input events (Figure 6.2).

It is normal to use Boolean algebra symbols to indicate the type of gate and the trees are usually presented from the top of a sheet downwards. Some analysts, however, work from left to right and use the notation shown in Figure 6.3.

Figure 6.4 on page 84 provides a list of standard Boolean symbols.

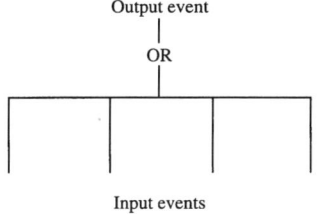

Figure 6.1   'OR' gate.

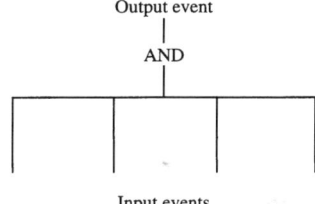

Figure 6.2   'AND' gate.

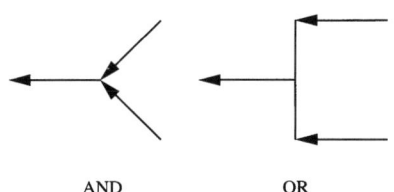

Figure 6.3   Alternative symbols.

83

LOGIC SYMBOLS

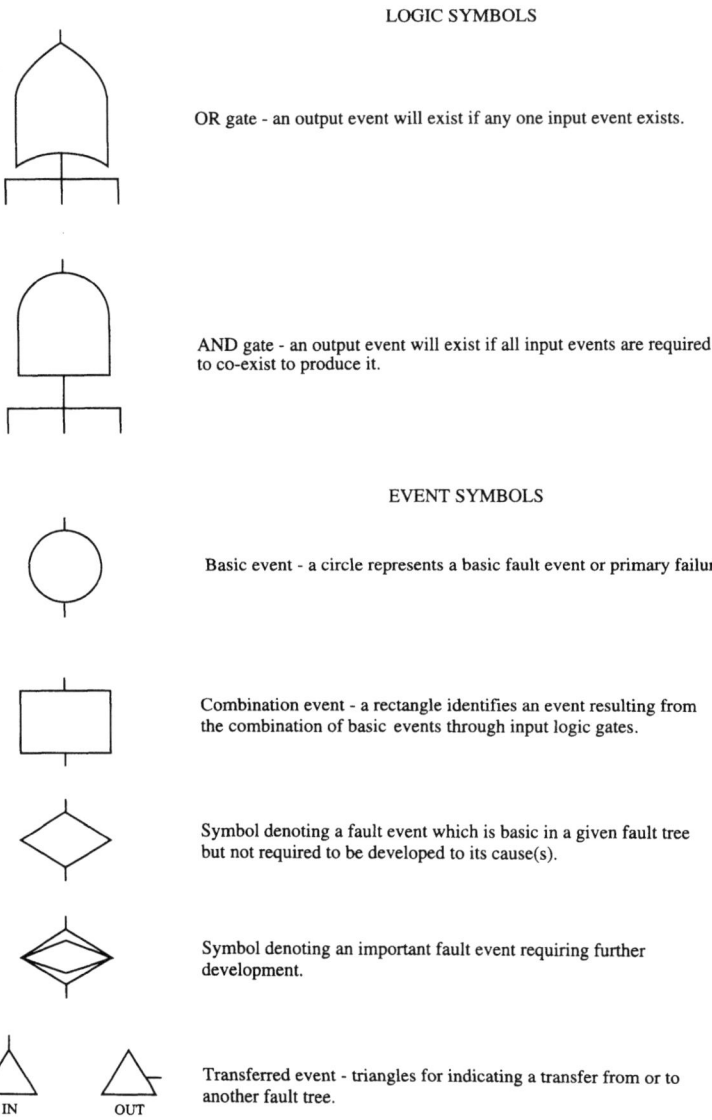

OR gate - an output event will exist if any one input event exists.

AND gate - an output event will exist if all input events are required to co-exist to produce it.

EVENT SYMBOLS

Basic event - a circle represents a basic fault event or primary failure.

Combination event - a rectangle identifies an event resulting from the combination of basic events through input logic gates.

Symbol denoting a fault event which is basic in a given fault tree but not required to be developed to its cause(s).

Symbol denoting an important fault event requiring further development.

Transferred event - triangles for indicating a transfer from or to another fault tree.

IN   OUT

Figure 6.4   Standard symbols used in Fault Tree Analysis.

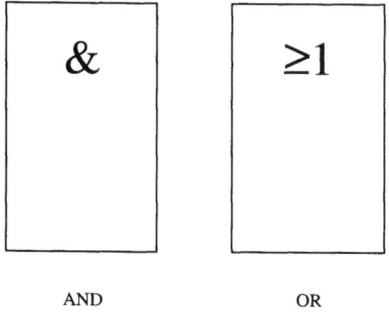

AND   OR

Figure 6.5   IEC 1025 symbols.

A further form of notation is recommended in BS 5760 (IEC 1025)[1]; in this case the top event is placed at the left hand side of the diagram and the symbols for the AND and OR gates are shown in Figure 6.5. The other symbols are as shown in Figure 6.4.

The technique of Fault Tree Analysis can be illustrated by a simple non-technical example, reproduced with the permission of AEA Technology plc. Figure 6.6 (page 86) shows the exterior of a typical English public house with one traveller arriving by car and others arriving on foot and by bicycle. The public house has a car park and two bars, public and lounge, each with its own entrance. Figure 6.7 (page 86) shows the interior with the two bars, each with their own beer supply system and bar staff. Note that the two systems are totally separate and identical except for the staff.

As with all safety analysis it is first necessary to define the scope of the analysis. In this case the undesired event is 'no beer for the car traveller'. Figure 6.8 (page 87) shows the unquantified fault tree for this event. The use of the AND and OR gate logic is clearly shown. Note the use of two parallel routes once the traveller has been able to park his car.

Figure 6.9 (page 88) shows a further very simple unquantified fault tree drawn left to right using the alternative symbols favoured by some analysts in the process industries.

Figure 6.10 (page 89) shows a generic fault tree indicating the basic logic construction for a typical process plant. Thus an incident will only occur if a dangerous situation has been allowed to develop, the operators have been unable to take mitigating action and the final trips or protective devices have also failed.

Figure 6.6   Exterior of public house. © AEA Technology plc 1996

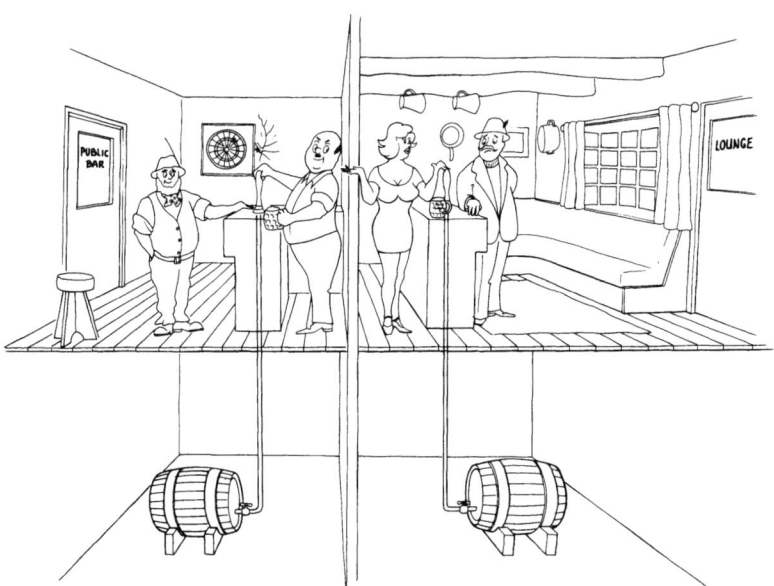

Figure 6.7   Interior of public house. © AEA Technology plc 1996

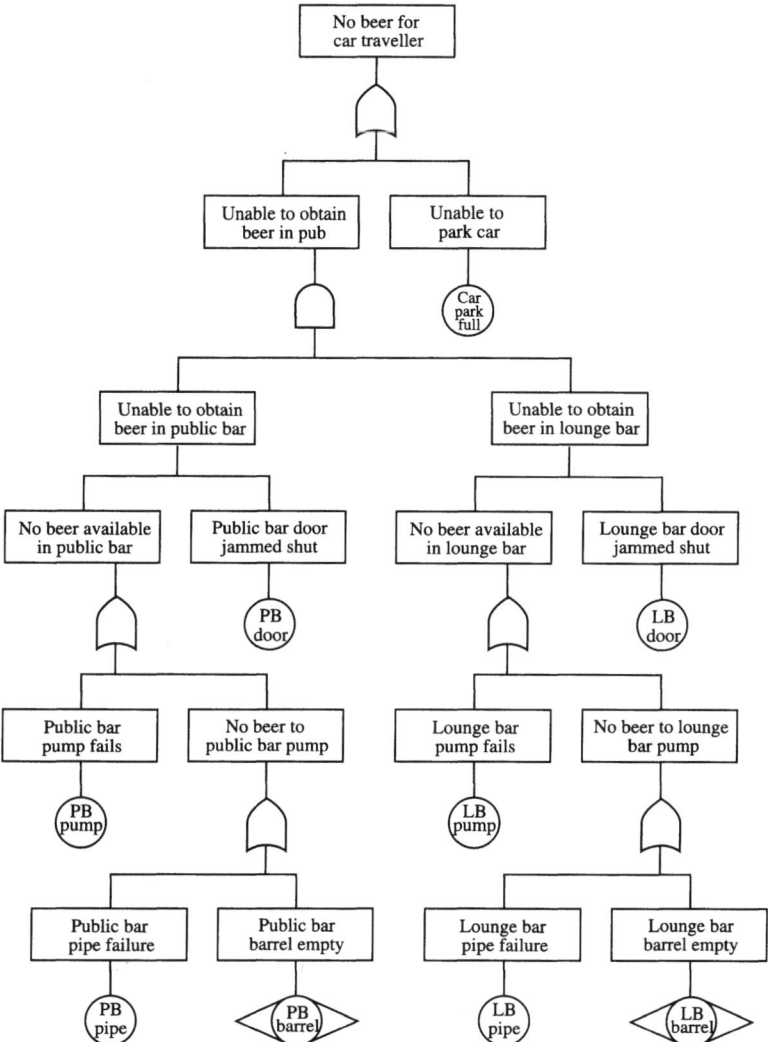

Figure 6.8   Public house fault tree.

COMMON EVENTS

These examples show three types of common event. There is a considerable amount of confusion in the terms used to define common events and the litera-ture is by no means consistent. A number of common events or situations can be identified which affect the way in which a tree is constructed and quantified,

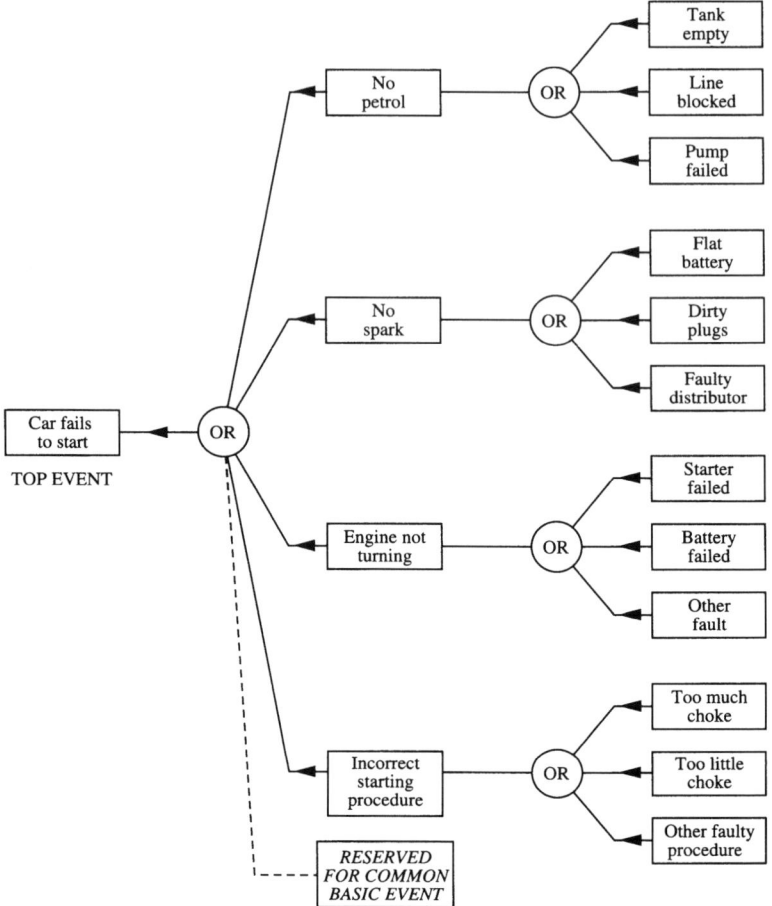

Figure 6.9   Failure of a car to start.

all requiring different treatments. It is therefore important to have a clear understanding of these effects and how best to handle them.

Common basic event
The same basic event may appear in more than one branch of a tree — for example, a flat car battery may prevent a car starting either because of lack of
power to the starter motor or lack of energy to the spark plugs, as shown in Figure 6.9. Provided that such events are identified at the time and the rules of
Boolean algebra applied, the correct result will be achieved. Alternatively such
events can be taken out of the individual sub-trees and shown as a basic event

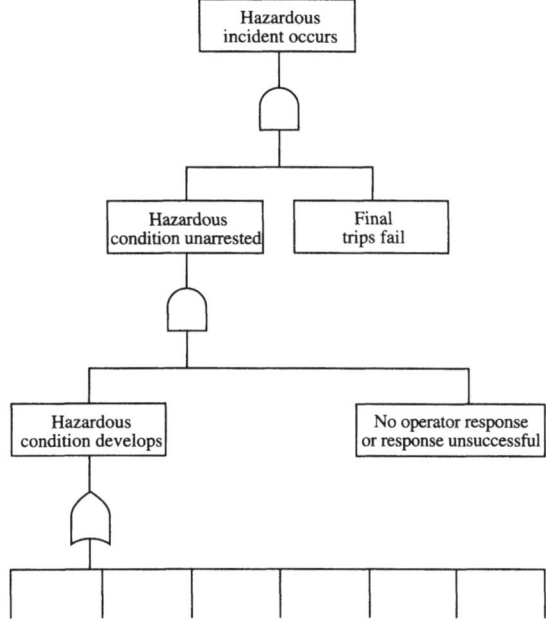

Base events leading to hazardous condition

Figure 6.10   Basic fault tree.

as indicated in Figure 6.9. The main problem is often one of early identification, as the example of the car shows.

Common cause

A number of components or subsystems can fail due to the same cause — for example, flood, fire, vibration and so on — and this must be identified in the analysis. In the case of the public house, the cellar could be flooded or damaged by fire. Whilst the fault tree can be a useful technique to identify such problems, they are best solved by design. Thus trip systems should be located in separate fire or flood compartments and duplicate power and signal lines should be physically separated as far as possible — two metres is usually accepted as adequate. In the public house example it can be seen that the two cellars are clearly separated by a partition wall.

Common mode

A system may well contain a number of identical components — for example, the whole beer supply system in Figure 6.8. The components may perhaps be to the same design, from the same manufacturing batch and installed by the same technicians. In this case there is always the risk of generic faults due to design, manufacture or installation. Thus the degree of protection provided by duplicate systems is much reduced.

The analysis of such failures is complex and a simplified treatment is discussed in Section 7.3 on page 117. The best method of protection is to avoid the situation by design. Thus a trip may be initiated by a temperature and a pressure device rather then by, say, two temperature detectors. If this approach is not practical then it may be possible to use two different types of sensor — for example, a thermocouple and a resistance thermometer. The use of two different designs or manufacturers of the same type of device will give a considerable enhancement of protection. The public house landlord should consider the use of different designs of beer pump in the two bars.

The terms 'common mode' and 'common cause' are closely inter-linked and are often used for the same thing. The term 'dependent failure' is also used by some analysts to describe this situation.

BASIC RULES FOR LOGIC TREE CONSTRUCTION
The application of the following basic rules should ensure a soundly-based tree in a form suitable for quantification:
- think one step at a time in the correct sequential order;
- define the top event, then think of the essential prerequisites for that event;
- inter-relate them with the top event by AND or OR gates as appropriate;
- sub-divide complex systems into sub-trees;
- continue downwards through the sub-trees until the base or initiating events are identified;
- check for all categories of common events.

6.4   PROBABILITY THEORY
Two main aspects of probability have to be considered, one being time independent and the other time dependent:
- the probability of a particular outcome from a specified event — for example, success or failure when performing an action on a specific occasion;
- the probability of a particular state or condition existing in a set time domain — for example, the availability of a piece of equipment during a given time period.

In both cases probability is a dimensionless number between zero and unity, zero representing absolute impossibility and unity representing absolute certainty.

$$0 \leq P \leq 1$$

It follows that the probability of a particular outcome (or a particular state existing), P, plus the probability of it not occurring, $\overline{P}$, must be 1.

$$P + \overline{P} = 1 \quad \text{and} \quad P = 1 - \overline{P}$$

In the case of discrete events — for example, the throw of a dice — the probability can be derived on the basis of straight chance or on the basis of observed results. Thus the probability of throwing a six is $\frac{1}{6}$ and if n missions out of a total of M were successful then the probability of a successful mission is $\frac{n}{M}$. It may also be decided on the basis of experience or judgement; if a plant operator is likely to open the wrong valve in one out of every 200 operations, then the failure probability would be 0.005.

COMBINING PROBABILITIES

Consider two entirely separate events which can lead to outcomes A and B respectively which are completely independent of each other. If the probabilities of outcomes A and B are given by P_A and P_B then the probability of outcome A AND outcome B is given by:

$$P_{\text{A and B}} = P_A \times P_B = P_A P_B$$

The probability of either outcome A OR outcome B OR both is given by:

$$
\begin{aligned}
P_{\text{A or B}} &= P_A \overline{P_B} + P_B \overline{P_A} + P_A P_B \\
&= P_A(1 - P_B) + P_B(1 - P_A) + P_A P_B \\
&= P_A + P_B - P_A P_B \\
&= P_A + P_B \quad \text{if } P_A \text{ and } P_B \text{ are small.}
\end{aligned}
$$

Similarly it can be shown that:

$$
\begin{aligned}
P_{\text{A and B and C}} &= P_A P_B P_C \\
P_{\text{A or B or C}} &= 1 - \overline{P_A P_B P_C} \\
&= P_A + P_B + P_C \quad \text{if } P_A \text{ and } P_B \text{ and } P_C \text{ are small.} \\
P_{\text{total}} &= 1 - \prod_i (1 - \overline{P_i})
\end{aligned}
$$

Though failure probabilities are usually low, always check validity when using the above approximations.

6.5   SET THEORY AND BOOLEAN ALGEBRA

The theory of Boolean algebra was developed in the middle of the last century by George Boole, a British mathematician. It was reformulated as the theory of sets towards the end of the century but it was not until the advent of electronic circuitry that it found a practical application. Because of the history of its development, three sets of notation are found in the literature. For the purposes of quantitative safety analysis, the form used in electronics and computing is usually employed.

SET THEORY
A set is best expressed as a list either explicitly or implicitly. For example:

$A = \{1, 2, 3, 4, 5, 6, 7, 8, 9\}$
$B = \{2, 4, 6, 8, 10, 12\}$
$C = \{2, 4, 6, 8\}$
$D = \{\text{even numbers}\}$

Figure 6.11 gives a pictorial representation of a set — a Venn diagram. Sets are usually represented by upper case letters.

Two sets are equal if they contain the same elements. For example:

{James, Richard, Mary} = {Richard, Mary, James}

Membership of a set is expressed as $x \in M$ and non membership as $x \notin M$. Thus, for the example above, $1 \in A$ but $1 \notin B$.

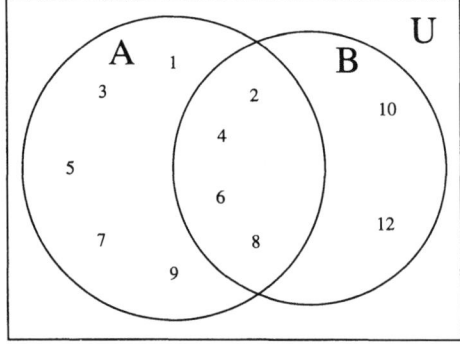

Figure 6.11   Venn diagram.

A set X is a subset of set Y if all the elements of X are also the elements of Y. This is expressed as $X \subset Y$. Thus, for the examples, $C \subset D$ and $C \subset A$.

Two special cases must be considered. The first is the set that contains no elements at all and is called a null set and denoted by 0 or ϕ. The other is the set which includes all objects under consideration and is called the universal set, denoted by 1 or U. The complementary set of A is a set that contains all the elements that do not belong to A and is denoted by $\overline{A}$ or A'.

Sets may also be defined in terms of other sets — for example, a set C could be defined as a set consisting of all the elements of set A and all the elements of set B and no others as shown below:

A = {1, 2, 3}
B = {1, 4, 7}
C = {1, 2, 3, 4, 7}

It will be seen that the elements only appear once in set C which is defined as a union of sets A and B and denoted by $C = A \cup B$. This is usually simplified to $C = A + B$.

Similarly a set C can be defined that contains only those elements which appear in both sets A and B. For example:

A = {1, 2, 3}
B = {1, 4, 7}
C = {1}

This set is defined as the intersection of sets A and B and is denoted by:

$C = A \cap B$, usually simplified to $C = A \cdot B$.

APPLICATION TO PROBABILITY

Boolean algebra can be applied to probability theory because we can associate the outcomes of trials with events. Thus if the trial is the throw of a die then the event A1 can be defined as:

A1 = {lands on 1}

Thus the universal set would be:

U = {lands on 1, lands on 2, lands on 3, lands on 4, lands on 5, lands on 6}

and a null set would be:

$\phi = \{\text{lands on 7}\}$

If an experiment is performed and the event A occurs then it can be expressed as:

$A = 1$

and if A does not occur then:

$A = 0 \text{ or } \overline{A} = 1$

$P(A = 1)$ is the probability that A will be the result.

BOOLEAN MANIPULATION

In order to be able to use Boolean algebra for the evaluation of fault trees, it is necessary to understand the basic rules of Boolean manipulation. The order of evaluation of a Boolean expression follows the same rules as ordinary algebra. Thus to evaluate $A + B \cdot C$ it is first necessary to evaluate $B \cdot C$, but if the expression were written as $(A + B) \cdot C$ then $A + B$ would be evaluated first. The following rules of ordinary algebra also apply:

- both operations are commutative:

$A + B = B + A \text{ and } A \cdot B = B \cdot A$

- both operations are associative:

$(A + B) + C = A + (B + C) \text{ and } (A \cdot B) \cdot C = A \cdot (B \cdot C)$

- the multiplicative distributive law holds:

$(A + B) \cdot (C + D) = A \cdot C + B \cdot C + A \cdot D + B \cdot D$

- both operations have identities:

$A + 0 = A \text{ and } A \cdot 1 = A$

The differences are equally important:
- the complementary set:

$A + \overline{A} = 1 \text{ (that is, either A or A must occur)}$

$A \cdot \overline{A} = 0 \text{ (that is, A and A cannot co-exist)}$

Complementation operations are performed before multiplication and addition unless overridden by a bracket.

- the idempotent law:

$$A + A = A \text{ and } A \cdot A = A$$

- the absorption law:

$$A + A \cdot B = A$$

- De Morgans Laws

$$\overline{(A + B)} = \overline{A} \cdot \overline{B}$$
$$\overline{(A \cdot B)} = \overline{A} + \overline{B}$$

APPLICATION TO FAULT TREES

In quantitative safety analysis using logic or fault trees two types of event are considered — basic events and combination events, The top event is the final combination event. In the simple example in Figure 6.12, A, B and C are basic events, Q and R are intermediate combination events and P is the final combination or top event.

Thus: $Q = A + B$
$R = A + C$
$P = Q \cdot R$

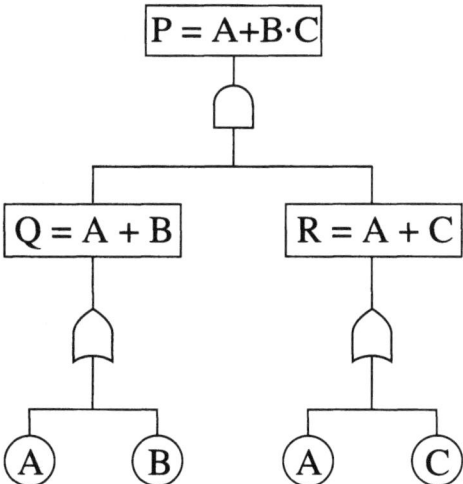

Figure 6.12 Simple fault tree.

An AND gate is a Boolean product and an OR gate is a sum. Applying the laws of Boolean algebra gives:

$$
\begin{aligned}
\text{Top } (P) &= (A + B) \cdot (A + C) \\
&= A \cdot A + A \cdot C + B \cdot A + B \cdot C \\
&= A + A \cdot C + B \cdot A + B \cdot C & (A \cdot A = A) \\
&= A + B \cdot A + B \cdot C & (A + A \cdot C = A) \\
&= A + B \cdot C & (A + B \cdot A = A)
\end{aligned}
$$

CUT SETS

The use of cut sets can assist in the evaluation of fault trees. A cut set is defined as a combination of basic events, the simultaneous existence of which will produce the top event. A minimal cut set is defined as a cut set that has no other cut set as a proper subset — that is, if any basic event is removed from the set the top event will not occur.

In the above example A, AC, BC and ABC are all cut sets but only A and BC are minimal cut sets as AC and ABC have A as a subset.

Most computer-based methods of evaluating fault trees are based on these principles.

6.6   COMBINATION OF FREQUENCIES

Assume that two totally independent systems A and B can lead to the potentially dangerous recurring events A and B which, if they occur simultaneously, can give rise to a dangerous condition AB. Event A occurs at a frequency of λ_A, event B occurs at a frequency of λ_B per year and the duration of the two events are D_A and D_B yr respectively.

Provided that λ_A is very small, the probability that event A is taking place is given by:

$$ P_A = \lambda_A D_A $$

Similarly:

$$ P_B = \lambda_B D_B $$

A dangerous condition will arise if events A and B overlap — that is, if event B occurs whilst event A is already taking place (λ_{AB^*}) or if event A occurs whilst event B is already taking place (λ_{A^*B}). This can be expressed mathematically as follows:

$$ \lambda_{AB^*} = \lambda_B P_A = \lambda_B \lambda_A D_A $$

$$\lambda_{A^*B} = \lambda_A P_B = \lambda_A \lambda_B D_B$$

Thus the combined frequency of the two dangerous events is:

$$\lambda_{AB} = \lambda_{AB^*} + \lambda_{A^*B} = \lambda_B \lambda_A D_A + \lambda_A \lambda_B D_B$$
$$= \lambda_A \lambda_B (D_A + D_B)$$

If $D_A \gg D_B$ then $\lambda_{AB} = \lambda_A \lambda_B D_A$
If $D_A = D_B$ then $\lambda_{AB} = 2\lambda_A \lambda_B D_A$
If $\lambda_A D_A = 1$ then $\lambda_{AB} = \lambda_B$

DURATION OF COINCIDENCE OF EVENTS
The probability that the recurring events will coincide is given by:

$$P_{AB} = P_A P_B = (\lambda_A D_A)(\lambda_B D_B)$$

The average duration for coincidence of the two recurring events is:

$$D_{AB} = \frac{P_{AB}}{\lambda_{AB}} = \frac{(\lambda_A D_A)(\lambda_B D_B)}{\lambda_A \lambda_B (D_A + D_B)} = \frac{D_A D_B}{D_A + D_B}$$

This rule for the combination of frequencies is very important. It must always be remembered that:

$$\lambda_{A \text{ and } B} \neq \lambda_A . \lambda_B$$

EXAMPLE
Two relief valves discharge into a common header (Figure 6.13). The header has been undersized and could rupture if both valves discharge at the same time.

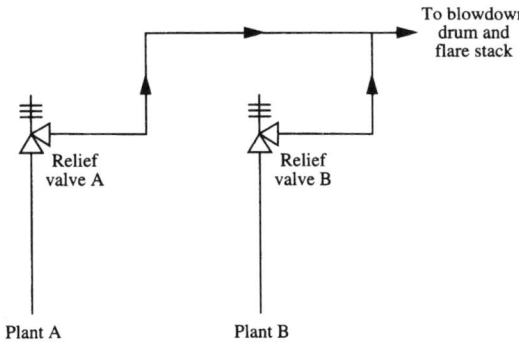

Figure 6.13 Relief header.

Using the data below, estimate the frequency for the rupture of the header and the duration of discharge from the ruptured header.

$$\lambda_A = 4 \text{ yr}^{-1}, \lambda_B = 1 \text{ yr}^{-1}$$

$$D_A = 0.75 \text{ hr}, D_B = 0.5 \text{ hr}$$

$$\lambda_{AB} = \lambda_A \lambda_B (D_A + D_B) = 4 \times 1(0.75 + 0.5)/8760 = 0.00057 \text{ yr}^{-1}$$

Thus the frequency of rupture is about 0.0006 per year.

The duration of emission is given by:

$$D_{AB} = \frac{D_A D_B}{D_A + D_B} = \frac{0.75 \times 0.5}{0.75 + 0.5} = 0.3 \text{ hr}$$

6.7   CONCLUSIONS

The basic rules for the combination of probabilities and frequencies can be summarized as follows:

$$\begin{aligned}
P_{A \text{ and } B} &= P_A \cdot P_B \\
P_{A \text{ or } B} &= P_A + P_B - P_A P_B \\
&= P_A + P_B \text{ if } P_A \text{ and } P_B \text{ are small} \\
\lambda_{A \text{ and } B} &= \lambda_A \lambda_B (D_A + D_B) \\
\lambda_{A \text{ or } B} &= \lambda_A + \lambda_B \\
\lambda_{A \text{ and } B} &\neq \lambda_A \cdot \lambda_B
\end{aligned}$$

It is permissible to multiply frequencies by probabilities; this point is expanded further in Section 7.2 (page 112) in connection with fractional dead time.

6.8   EVENT TREE ANALYSIS (ETA)

So far only the backward reasoning or deductive approach to quantitative safety has been considered, in which the logic goes in the direction of effect to cause. There is also a need for an inductive approach which works from cause to effect; the former identifies the basic cause, the latter the ultimate consequences.

This technique is called Event Tree Analysis (ETA). It was first developed in the aerospace and defence industries but has now spread to the chemical and nuclear industries. Event trees are basically a modified form of the logic trees used in decision analysis and describe possible progressions from a given initiating event via a series of binary branches representing success or failure of safety features or the occurrence or not of some physical event. Thus the event tree is a diagrammatic representation of all the potential ways in which an event (the initiating event) can develop through a system.

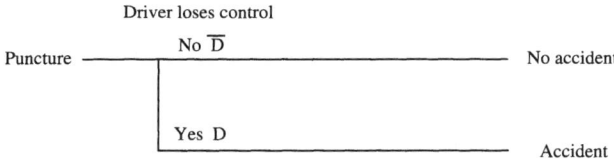

Figure 6.14   Event tree for puncture.

NOTATION

The approach in this case is left to right rather than top to bottom with the initiating event on the left and developing consequences expanding to the right. This is best illustrated by a simple example.

A car is being driven along a road when one of the front tyres punctures (the initiating event). If the driver manages to control the car there will be no accident; if he does not there will be an accident. This is expressed as the event tree shown in Figure 6.14.

The branch point is where the logic path splits at the nodal question/event — does the driver lose control? Each path or accident sequence is assigned a system outcome — that is, accident or no accident.

The convention used in event trees always puts the mitigating or less damaging event in the upper branch. Thus the tree shows an increase in the severity of the consequence as it goes down the page.

Each event can be assigned a Boolean variable — for example:

$D = \{\text{Driver loses control}\}$

so the upper path is represented by the Boolean event $\bar{D}$ and the lower path by D.

Again the convention places the unbarred or failed variable in the lower branch and the successful or barred variable in the upper branch.

Figure 6.15 (page 100) shows a slightly more complex example with a total of five paths. The top path which has no branches shows that the effectiveness or otherwise of subsequent nodes — for example, 'curtain fire extinguished' — are irrelevant to this branch and are said to be straight-lined at this point.

Using Boolean notation the paths can be described as follows:

Path 1 = $\bar{D}$
Path 2 = $D\bar{C}$
Path 3 = $DC\bar{R}$
Path 4 = $DCR\bar{H}$
Path 5 = $DCRH$

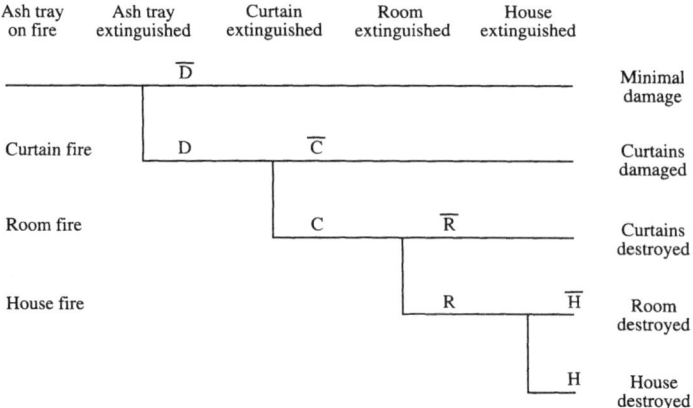

Figure 6.15 Event tree for cigarette fire.

Probabilities can thus be applied to each branch point, and multiplying the probabilities along the path yields the probability of occurrence of that path. By logic and the rules of Boolean algebra the total of all paths must be unity.

EVENT TREE CONSTRUCTION

Assuming the initiating event has been identified, the analyst must determine the nodal events and arrange them in order. The nodal events may be engineered safety features — such as fire extinguishers as in the above example — or they may be physical events. The analyst must have a good knowledge of the system in order to be able to produce the right list of events. The events are placed in a logical order; often this can be done chronologically on the basis of the time at which each nodal question is asked. It is necessary to ensure that all prerequisite questions regarding a particular nodal point have already been asked. Thus in the example the question 'has the curtain fire been extinguished?' must be asked before questioning the room fire.

Once the order has been determined the tree can be drawn out. It is necessary to decide at each branch point whether or not to straight-line the node. A node can be straight-lined if there is no need for the subsequent nodes to function. So in the example if the fire is immediately detected and extinguished there is no need for the other nodes. A further condition, not illustrated, is where the severity of the accident progression is such that success at this branch point cannot arrest it under any circumstances.

When the tree is complete it must be quantified by adding probabilities to each branch point. The sources of data are the same as those for other forms of quantitative safety analysis. Human error data is frequently needed in event trees and the analyst's judgement is very important.

The main steps in the creation of an event tree can be summarized as:

- categorize initiating event;
- determine functional dependencies;
- produce a logic tree;
- assign probabilities of failure to each node;
- calculate branch probabilities;
- review reasonableness of results.

PROBLEM AREAS

The main problem areas are similar to other forms of QRA — the difficulty of ensuring that the logic tree is correct and of obtaining reliable data. It is important to have some idea of the levels of confidence in the result, and the use of sensitivity analysis can help here. Because the method depends very much on the analyst's judgement, a rigorous quality assurance system involving peer review and checking is essential.

ADVANTAGES OF ETA

- Probabilities can be assigned to nodes on a path-dependent basis.
- A whole range of outcomes can be accommodated in one tree (a fault tree can only discriminate between success and failure).
- An event tree moves from cause to effect and is an easier concept to understand.
- Event trees are time-dependent.
- Event trees are ideal in examining the response of systems and people to a potential accident.

DISADVANTAGES OF ETA

- Event trees grow very quickly so, in order to keep to a reasonable size, the analysis is of necessity very coarse.
- The mathematics lacks the sophistication of that used in FTA.

6.9   CONCLUSIONS

The logic tree approach enables the various factors which can cause an incident, or effect its outcome, to be combined in a manner suitable for quantification. Fault trees examine the ways in which the failure of different parts of a system and/or human error can interact to give rise to an undesired event. Event trees can then be used to examine the ways in which the situation is likely to develop. The data required to quantify the trees are discussed in Chapter 7.

REFERENCES IN CHAPTER 6

1.   BS 5760, Part 7, 1991, Guide to fault tree analysis (BSI, UK) (IEC 1025) 1990.

FURTHER READING

1.   Kletz, T.A., 1992, *Hazop and Hazan: Identifying and Assessing Process Industry Hazards*, 3rd edition (IChemE, Rugby, UK).

7. QUANTIFICATION OF LOGIC TREES

7.1   INTRODUCTION

Data on the reliability of the components and systems involved is needed before a logic tree can be quantified. Every piece of equipment has its own inherent failure rate and the reasons for failure can be many and varied. Equipment can also be over-stressed or subjected to environmental conditions that are outside its design limits.

It is commonly accepted that the failure data for most equipment is based upon a lifetime characteristic often known as the 'bath-tub' curve (see Figure 7.1).

The failure rate changes with the life of the system or component and passes through three distinct phases:

(1) Inherent fault or wear-in period. This covers the 'running in' period during which design, material and installation faults manifest themselves. This is often called the 'infant mortality' period.

(2) Useful life period. This is the period of greatest interest in QRA. During this period faults are unrelated to age and occur randomly. The failure rate can be considered constant.

(3) Wear-out period. In this period failure is due to the deterioration caused by age or usage.

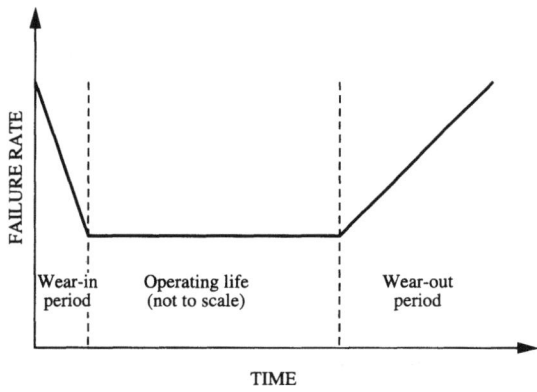

Figure 7.1   The 'bath-tub' curve.

SOURCES OF DATA

Failure rate data may be obtained in a number of ways:

- by sample testing, a method suitable for mechanical and electrical components. Component data may at times be combined to give system data. One limitation in its application to process plant is that the data are specific to the environment in which the tests are carried out;
- from data banks, some of which are available in the public domain and others by subscription. Much of the information in data banks comes from the nuclear and defence industries where extremely high standards of quality assurance have always been demanded. This means that some of the data might be rather optimistic for use in the general process industries. As with sample testing, it is important to ensure that environmental conditions are taken into consideration. References 1, 2 and 3 are good sources of data for process use;
- from plant experience. This is the best source of data but it has taken many organizations a long time to realize that they have such information available in the form of maintenance records. Modifications are usually needed to normal company maintenance record-keeping if the data are to be useful to risk analysts;
- by predictive techniques, making use of the fact that data may be available on the constituent parts of a system. Provided that the system is properly analysed, such data can be combined to produce an overall failure rate.

A selection of data sources with particular applicability to the chemical and process industries is given in Table 7.1. Full details of these and many other sources of data are given in Reference 3.

TABLE 7.1
Sources of failure data

Title	Publisher	Format	Date
CCPS	AIChE	Book	1990
ENI Reliability data book	ENI (Italy)	Book	1982
NPRD 91	Reliability Analysis Centre, New York	Database	1991
OREDA 92	DNV Technica	Book	1992
Davenport & Warwick	UKAEA (SRD)	Report	1991
WASH 1400	US NRC	Report	1974
SRD Data Centre	AEA Technology	Database	Ongoing
RMC Haris	RM Consultants	Database	1992

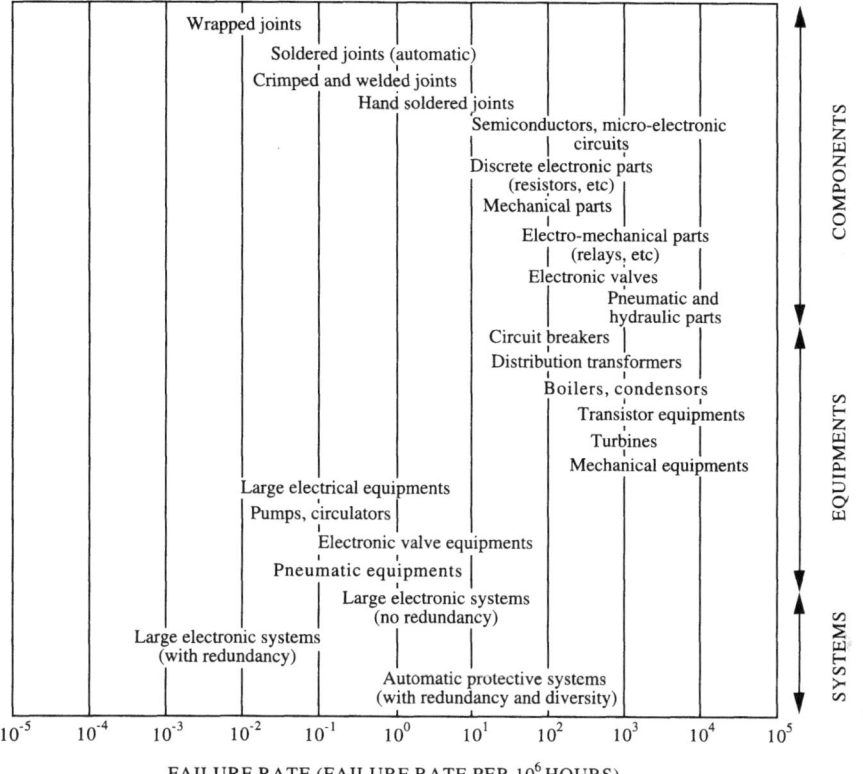

FAILURE RATE (FAILURE RATE PER 10^6 HOURS)

Figure 7.2   Failure rate changes.
Reprinted by permission of John Wiley & Sons, Ltd.

Figure 7.2 (from Reference 3) shows the range of failure rates for different types of components and systems. The wide ranges for many items should be noted; they are due to such factors as intended use, levels of quality assurance and so on. Thus a Class 1 welded pressure vessel will have a much lower failure rate than a site-fabricated static water tank. Table 7.2 on page 106 (based on Reference 3) gives some information on process plant failure rates. Tables 7.3 and 7.4 on page 107 (also based on Reference 3) give factors to show how generic data must be modified to take environmental and stress factors into consideration.

Figure 7.3 (page 108) shows a cooling subsystem and Table 7.5 (page 108) shows how the failure rate for the subsystem can be predicted from published data[3].

TABLE 7.2
Process equipment failure rates[†]

Item	Failure rate*
Air supply	8
Alarm system	24
Blower	185
Boiler (steam)	220
Compressor (air)	51
Compressor (large centrifugal)	4000
Compressor (large reciprocating)	8000
Controller (flow)	30
Fan	50
Generator (diesel)	900
Heat exchanger (shell and tube)	40
Motor (electric)	8
Power supply	32
Pump (centrifugal)	200
Pump (fire, engine driven)	500
Tank	4
Turbine (steam)	35
Valve (pneumatic)	60
Valve (safety)	30
Vessel (pressure)	1

[†] Based on material from *Reliability of Mechanical Systems*, 2nd edition, 1994, by permission of Mechanical Engineering Publications.

* All failure rates are per million hours

Much of the available failure data simply gives overall failure rates and it is important in some cases to be able to apportion this data to particular modes. An indication of the available data[3] and the way in which it can be used is given in Table 7.6 (page 109). This information does not give the full picture because it is also important to know how each of the above failure modes contributes to

TABLE 7.3
Environmental stress factors[†]

Environmental condition	Factor
Ideal static	0.1
Vibration-free, controlled environment	0.5
General purpose, land-based	1.0
Ship	2.0
Road	3.0
Rail	4.0
Air	10.0

[†] Based on material from *Reliability of Mechanical Systems*, 2nd edition, 1994, by permission of Mechanical Engineering Publications.

TABLE 7.4
Rating stress factors[†]

Percentage of normal rating	Factor
140	4.0
120	2.0
100	1.0
80	0.6
60	0.3
40	0.2

[†] Based on material from *Reliability of Mechanical Systems*, 2nd edition, 1994, by permission of Mechanical Engineering Publications.

the failure under consideration. If the relevant failure mode for a control valve is 'fail open' then the failure rate can be estimated as shown in Table 7.7 on page 109.

Some degree of judgement is needed on the part of the analyst in assessing the contributions of the various failure modes.

107

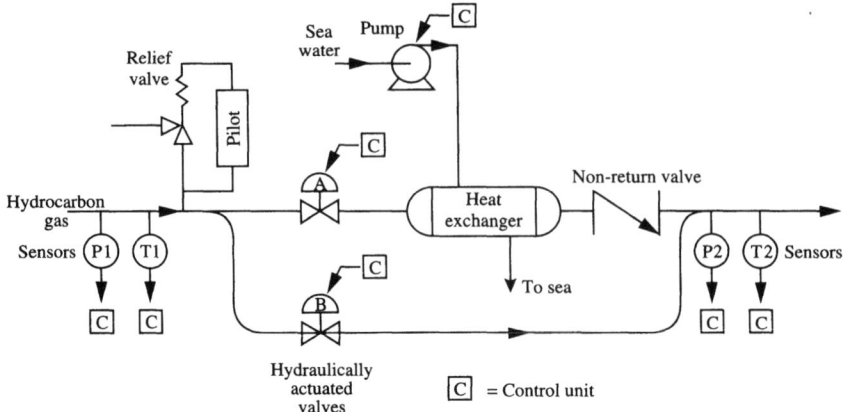

Figure 7.3   Cooling sub-system.
Reproduced with permission of Mechanical Engineering Publications, from *Reliability of Mechanical Systems*, 2nd edition, 1994.

TABLE 7.5
Failure rates for cooling subsystem†

Item	Mean failure rate*	No off	Total*
Pressure sensor	42	2	84
Temperature sensor	12	2	24
Safety relief valve	130	1	130
Shut-off valve	95	2	190
Heat exchanger	99	1	99
Check valve (NRV)	2	1	2
Centrifugal pump	260	1	260
Control unit	25	1	25
TOTAL			814*

† Based on material from *Reliability of Mechanical Systems*, 2nd edition, 1994, by permission of Mechanical Engineering Publications.

* All failure rates are per million hours

TABLE 7.6
Failure mode data†

Item	Failure mode	Proportion
Control valve	Leaks	50%
	Actuator	30%
	Sticks	10%
	Others	10%
Centrifugal pump	Critical	40%
	Degraded output	13%
	Incipient	45%
	Unknown	2%

† Based on material from *Reliability of Mechanical Systems*, 2nd edition, 1994, by permission of Mechanical Engineering Publications.

TABLE 7.7
Apportioning of failure mode†

Failure mode	Proportion	Contribution	Factor
Leaks	50%	10%	0.05
Actuator	30%	100%	0.3
Sticks	10%	100%	0.1
Others	10%	50%	0.05
TOTAL			0.5

Basic failure rate = 35 per 10^6 hours
Estimated FAIL OPEN rate = $35 \times 0.5 = 17.5$ per 10^6 hours

† Based on material from *Reliability of Mechanical Systems*, 2nd edition, 1994, by permission of Mechanical Engineering Publications.

ANALYSIS OF FAILURE DATA

Failure rate data may be expressed in a number of ways. The two most common are the number of failures per year and the number of failures per 10^6 hours. The time unit may be elapsed hours or operational hours, depending on the component and the circumstances. Other data may be expressed in terms of cycles, quantity of material processed, number of kilometres travelled and so on.

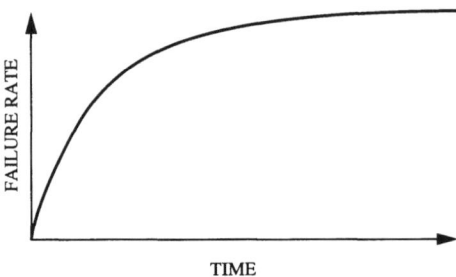

Figure 7.4   Exponential distribution.

It is important to obtain an indication of the probability that a component will fail in a given time, t, and this is termed the failure distribution function $F(t)$. Alternatively, analysts may be more interested in the probability of the component surviving for a given time, termed the reliability function $R(t)$. It is obvious that:

$$R(t) = 1 - F(t)$$

Various statistical distributions can be considered in the analysis of failure rate data. For most engineering components in the flat range of the 'bathtub curve' it has been shown that the most appropriate is the exponential distribution (Figure 7.4). For the exponential distribution the following equations apply:

$$F(t) = 1 - \exp(-\lambda t)$$

$$R(t) = \exp(-\lambda t)$$

Note: if $\lambda t \ll 1$ then $\exp(-\lambda t) = (1 - \lambda t)$ and $F(t) = \lambda t$

The probability of failure in time t is given by $P(t) = 1 - \exp(-\lambda t)$.

If the failure rate is low the probability of failure $P(t) = \lambda t$.

For more complex situations where the rate is not constant, the most frequently used model is the 'Weibull' distribution (see Figure 7.5). In the simple two parameter form $F(t)$ is given by:

$$F(t) = 1 - \exp\{(t/\eta)^\beta\}$$

where η is the scale parameter or characteristic life. It is the value of time t at which there is an approximately 2/3 probability that the component will have failed.

β is a shape parameter related to the failure rate. If $\beta < 1$, failure rate is decreasing; if $\beta = 1$, failure rate is constant and $\eta = 1/\lambda$ (exponential); if $\beta > 1$, failure rate is increasing.

REPAIRABLE COMPONENTS

Many components can be repaired and the simple theory assumes that the repair is as good as new — that is, the failure rate λ remains constant. If it is also assumed that the time to repair t_r is constant and that the defect is repaired as soon as it is discovered, it is possible to calculate the probability that the component will be working at any time t. The repair rate is expressed as:

$$\mu = 1/t_r$$

Therefore the probability that the component is not working in time period t is given by:

$$P(t) = \frac{\lambda}{\lambda + \mu} [1 - (\exp\{-(\lambda + \mu)t\}]$$

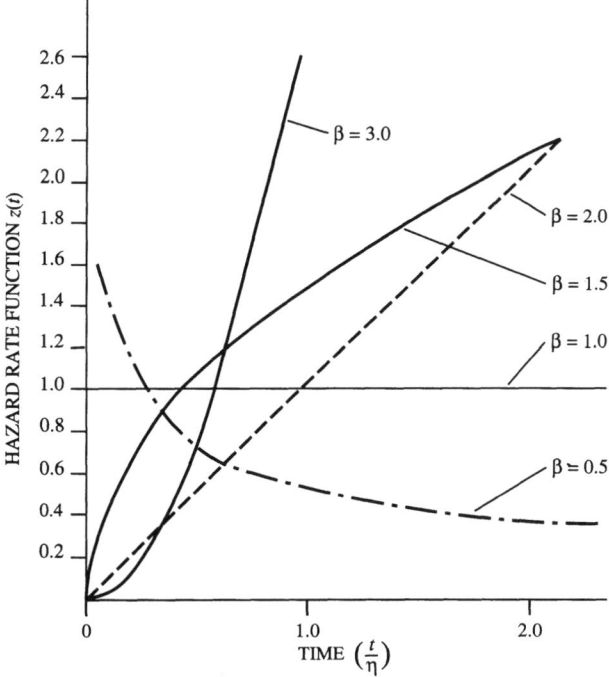

Figure 7.5   Weibull distribution.

111

7.2   FRACTIONAL DEAD TIME

Fractional dead time (*fdt* or ϕ, also known as mean unavailability) is the fraction of the total time that the protective device is in a failed state and is a measure of the probability of its failure to provide the protection intended by the design. If the *fdt* is zero then the device will always work when called upon, so the hazard rate will also be zero. If it is unity then the hazard rate will equal the demand rate — that is, it will provide no protection.

Thus hazard rate H = demand rate $D \times fdt$

$$H = \phi D$$

In the case of unrevealed failure, the operator may not know that a protective device is in a failed state until it is called upon to work. In order to reduce the *fdt* of a protective device, it is proof-tested at regular intervals to ensure that it is working or to reveal faults that can be repaired. The time between proof tests is termed the proof test interval, T_p.

Consider a protective device with an unrevealed fail to danger rate λ yr^{-1}. If it is proof-tested at intervals of T years there will be periods, as shown shaded in Figure 7.6, during which it will give no protection.

| 0 | T | 2T | 3T | 4T | 5T |

Figure 7.6   Proof test intervals.

The concept can be analysed as follows:

Device failure rate:	λ, yr^{-1}
Proof test interval:	T_p, yr
Total dead time:	T_d, yr
Fractional dead time (*fdt*):	ϕ
Demand rate:	D, yr^{-1}
Plant hazard rate:	H, yr^{-1}

Assuming exponential failure rate distribution, the probability of failure of the trip within a proof test period q is given by:

$$q = 1 - \exp(-\lambda T_{\mathrm{p}})$$

If $\lambda \ll 1$, $q = \lambda T_{\mathrm{p}}$

The trip is only required on demand; therefore the probability that demand will occur during the dead time is given by:

$$P_{\mathrm{d}} = 1 - \exp(-DT_{\mathrm{d}})$$

But on average $T_{\mathrm{d}} = T_{\mathrm{p}}/2$ (that is, trip will be dead for half of the total time between tests).

Hence $P_{\mathrm{d}} = 1 - \exp(-DT_{\mathrm{p}}/2)$

If $D \ll 1$ then $P_{\mathrm{d}} = DT_{\mathrm{p}}/2$

The probability that a plant hazard will occur during the proof test interval is given by:

$$P_{\mathrm{H}} = 1 - \exp(-HT_{\mathrm{p}})$$

If $H \ll 1$ then $P_{\mathrm{H}} = HT_{\mathrm{p}}$

But $P_{\mathrm{H}} = q \times P_{\mathrm{d}}$

(Probability of plant hazard = probability of trip failure × probability that demand will occur when trip is in failed state.)

If all the approximations apply then $P_{\mathrm{H}} = \lambda T_{\mathrm{p}} \times DT_{\mathrm{p}}/2$.

$$H = P_{\mathrm{H}}/T_{\mathrm{p}} = \lambda T_{\mathrm{p}} \times DT_{\mathrm{p}}/2T_{\mathrm{p}} = \lambda T_{\mathrm{p}}D/2$$

But $H = D\phi$ (demand rate × fractional dead time)

$\therefore \quad \phi = \lambda T_{\mathrm{p}}/2$

If the approximations do not apply, the hazard rate $H = \lambda P_{\mathrm{d}}$ (failure rate × probability that demand will occur during dead time)

$\therefore \quad H = \lambda\{1 - \exp(-DT_{\mathrm{p}}/2)\}$

Reducing the proof test interval is not totally beneficial because consideration must be given to the fact that the protective device may be disarmed during testing, and there is always a possibility of the device being rendered inoperative as a result of the test — for example, not being put back on line (see Figure 7.7 on page 114). The optimum test interval can be calculated by taking all these factors into consideration.

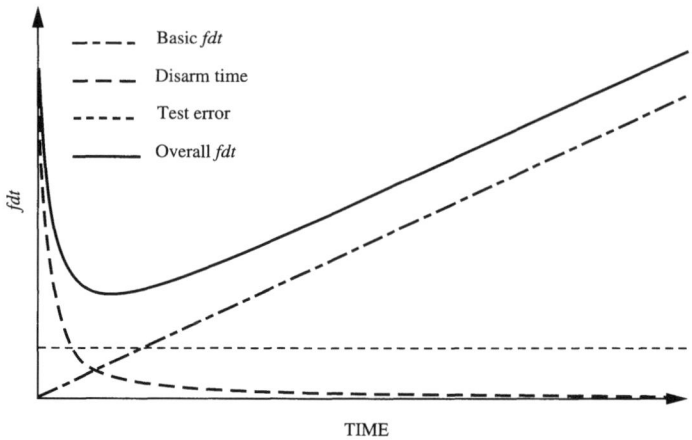

Figure 7.7   Fractional dead times.

EXAMPLE 1

A protective system has a failure rate of 0.01 yr^{-1} and is proof-tested at interval T yr. The system is off line for testing for a period of 1 hr (1/8760 yr) and the probability of a testing error is 1 in 1000 (0.001).

The total fractional dead time is given by:

$$\phi = 0.5 \times 0.01T + 1/8760T + 0.001$$

The minimum value is obtained when $\dfrac{d\phi}{dT} = 0$,

that is, when $0.005 - 1/8760T^2 = 0$, $T = 0.151$ yr (1.8 months).

$$\begin{aligned} \text{Thus } \phi &= 0.5 \times 0.01 \times 0.151 + 1/(8760 \times 0.151) + 0.001 \\ &= 2.51 \times 10^{-3} \end{aligned}$$

The values given by such calculations are often too low for practical purposes; test intervals of about six months are more common. This would give a value of $\phi = 3.73 \times 10^{-3}$.

If the demand rate is high relative to the proof test period then the device will be tested on demand rather than by proof test, so rendering the above approximation very inaccurate. This point can be illustrated with reference to a car braking system:

• failure rate $\lambda = 10^{-4}$ yr^{-1};

- test interval $t_p = 1$ yr (legal requirement);
- demand rate $D = 5 \times 10^4$ yr.

Using the simple formula:

$$\phi = 0.5 \times 10^{-4},$$

$$H = 0.5 \times 10^{-4} \times 5 \times 10^4 = 2.5 \text{ yr}^{-1}$$

This result is obviously wrong. Using the more exact formula:

$$H = \lambda\{1 - \exp(-DT_p/2)\}$$

$$H = \lambda = 10^{-4} \text{ yr}^{-1}$$

An approximate long time average formula to cover the situation where a device is tested by demand is:

$$H = \frac{\lambda D}{\lambda + D}$$

This formula is reasonably accurate when $D < \lambda$ but otherwise it tends to overestimate the hazard rate. If the demand rate D is much greater than the failure rate λ, it will be seen from the above equations that the hazard rate will equal the failure rate.

PARALLEL PROTECTIVE SYSTEMS

It is frequently considered necessary to improve safety and reliability by having two or more protective systems in parallel. The fractional dead time for such systems, assuming simultaneous testing, can be estimated as follows:

1 out of 2 $fdt = \dfrac{\lambda^2 T^2}{3} = \dfrac{4\,(fdt \text{ single channel})^2}{3}$

1 out of 3 $fdt = \dfrac{\lambda^3 T^3}{4}$

1 out of n $fdt = \dfrac{\lambda^n T^n}{(n + 1)}$

In order to reduce the problem of spurious or fail safe trips, a voting system may be used in which, say, two out of three protective systems must indicate a potentially hazardous condition before action is taken. In this case the fractional dead time is given by:

2 out of 3 $fdt = \lambda^2 T^2$

m out of n $\quad fdt = {}^{n}C_{r} \times \dfrac{\lambda^{r}T^{r}}{(r+1)}$

Where $r = n - m + 1$ and ${}^{n}C_{r}$ is the combination of n items taken r at a time:

$${}^{n}C_{r} = \dfrac{n!}{(n-r)r!}$$

A '1 out of n' system trip will operate if any one of the trips operates; an 'm out of n system' will only operate if m out of the n trips operate.

The fractional dead time of parallel protective systems may be reduced if the testing of the systems is staggered. In this way one of the trip systems is always active. The *fdt* of a two-channel system tested at intervals T with a test/repair time of t is given by:

$$(\textit{fdt of 1 out of 2}) \times \dfrac{T-2t}{T} + (\textit{fdt of 1 out of 1}) \times \dfrac{2t}{T}$$

Allowing for random test errors the average *fdt* of 1 out of 2 is given by:

Coincidence of random faults on both channels +

coincidence of random fault on A and test error on B +

coincidence of random fault on B and test error on A

The average *fdt* of 1 out of 1 is given by:

Random failure of that channel + test error

EXAMPLE 2

Assume that the system described in Example 1 on page 114 has two trips working on a '1 out of 2' basis. The testing is staggered at a total interval of T yr, each test taking t yr. The failure rate λ is 0.01 and the probability of test error P is 0.001. No allowance is required for disarm time as the other system will be operational during testing.

$$\phi_{(1 \text{ of } 1)} = 0.5\,\lambda T + P$$
$$= 0.005T + 0.001$$

$$\phi_{(1 \text{ of } 2)} = 1/3\,(\lambda T)^{2} + 2\,(0.5\,\lambda T)(P)$$
$$= 0.0000333T^{2} + 0.00001T$$

The fractional dead time of the combined system $\phi_{(\text{combined})}$ is given by:

$$[0.0000333T^{2} + 0.00001T]\dfrac{T-2t}{T} + [0.005T + 0.001]\dfrac{2t}{T}$$

Taking a test interval of 6 months (0.5 yr) and a test time of 1 hr (1.141×10^{-4} yr) the combined fractional dead time would be:

$$(1.333 \times 10^{-5}) \frac{0.5 - 2.282 \times 10^{-4}}{0.5} + (3.5 \times 10^{-3}) \frac{2.282 \times 10^{-4}}{0.5}$$

$$= (1.333 \times 10^{-5})(0.9995) + (3.5 \times 10^{-3})(4.566 \times 10^{-4})$$
$$= 1.332 \times 10^{-5} + 1.6 \times 10^{-6}$$
$$= 1.49 \times 10^{-5}$$

However, the question of dependent failure must also be considered.

7.3   DEPENDENT FAILURE ANALYSIS

It has been shown that a simple analysis will give over-optimistic results in cases where the cause of failure may be interdependent either by common cause or common mode. This factor is particularly important in the case of double or multiple redundant systems. There is unfortunately a lack of data on which to base the analysis of dependent failures and much of the data used was not collected with dependent failure analysis in mind. A number of methods of treating dependent failure have been published over the years but the β factor method is the one which has gained the highest degree of acceptance.

THE β FACTOR METHOD

This method was first published by Flemming[4] in 1974 in the USA. It is based on a series of assumptions:

(1) Failures in a redundant system may be of two types:
- failures that are completely independent — 'type 1';
- failures which are dependent or related — 'type 2'.

(2) If a type 2 failure occurs, all of the redundant components will fail (rather pessimistic).

(3) All redundant channels are identical and have the same failure rates.

The total failure rate of a redundant component is the sum of the failure rates for type 1 and type 2 failures:

$$\lambda = \lambda_1 + \lambda_2$$

$$\beta = \frac{\lambda_2}{\lambda_1 + \lambda_2} = \frac{\lambda_2}{\lambda}$$

$$\lambda_2 = \beta\lambda$$

The β factor is defined as the fraction of the total failure rate due to type 2 failures. The actual values of β have been the subject of much debate and depend on the degree of protection provided and the other factors discussed earlier. It is generally accepted that the values lie between 1 and 0.001 and the following figures are now generally accepted:
- identical redundancy 0.2
- partial diversity   0.06
- full diversity   0.02

Consider a two-unit identical redundant system, using an independent analysis. The system failure probability on demand is given by:

$$\frac{(\lambda T)^2}{3}$$

where λ is the failure rate and T the proof test interval. The dependent failure contribution is given by:

$$\frac{\beta \lambda T}{2}$$

Thus the total probability of failure is given by:

$$\frac{(\lambda T)^2}{3} + \frac{\beta \lambda T}{2}$$

EXAMPLE 3
Taking the data from Example 2, neglecting the probability of test errors and using a β factor of 0.2 gives:

$$\frac{(0.01 \times 0.5)^2}{3} + \frac{0.2 \times 0.01 \times 0.5}{2}$$
$$= 8.333 \times 10^{-6} + 5 \times 10^{-4}$$
$$= 5.083 \times 10^{-4}$$

Thus it is seen that the common mode effects predominate.

LIMITATIONS OF THE β METHOD
Although the method has the advantage of having only one parameter, the validity of some of the assumptions is doubtful. A whole range of extensions to the β method have been proposed to overcome the perceived weaknesses of the method. Most hinge around the concept of separating the coupling mechanisms which cause common mode and common cause failures from trigger or root causes of the failure. An analysis of such failures in US nuclear plants[4] has

shown the following range of contributions from the various coupling mechanisms that bring about dependent failure:

- same hardware/design   55%
- same staff (operation and maintenance) 25%
- same procedures   10%
- others (environment, location, etc)   10%

The field of dependent failure analysis is still the subject of much debate and a concerted effort is now being made to collect more data to substantiate the various methods of analysis. As with all methods of quantitative safety analysis, however, much always depends on the skill and judgement of the analyst.

7.4   WARNING

QRA is only as good as the data used; hence the importance of obtaining the best possible data. All sources of data should be carefully checked for applicability to the particular application. Protective systems must be analysed to determine the fractional dead time and particular attention paid to the potential for dependent failure modes.

7.5   APPLICATION OF FAILURE DATA TO FAULT TREES

Having produced the basic logic trees, the next steps are to obtain the necessary data for the basic events and then to combine the data through the logic gates using the rules described earlier in this chapter. This is best illustrated by a series of simple examples.

EXAMPLE 4

Figure 7.8 shows a simple flow control and trip system. The flow controller FC takes a signal from the orifice plate and transmitter FT and controls the flow of

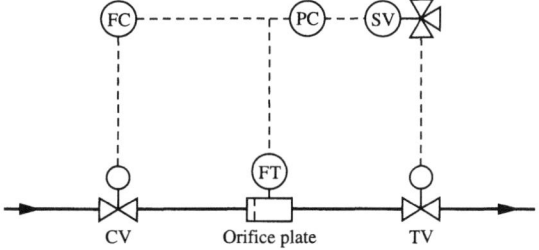

Figure 7.8   Flow control and trip system.

119

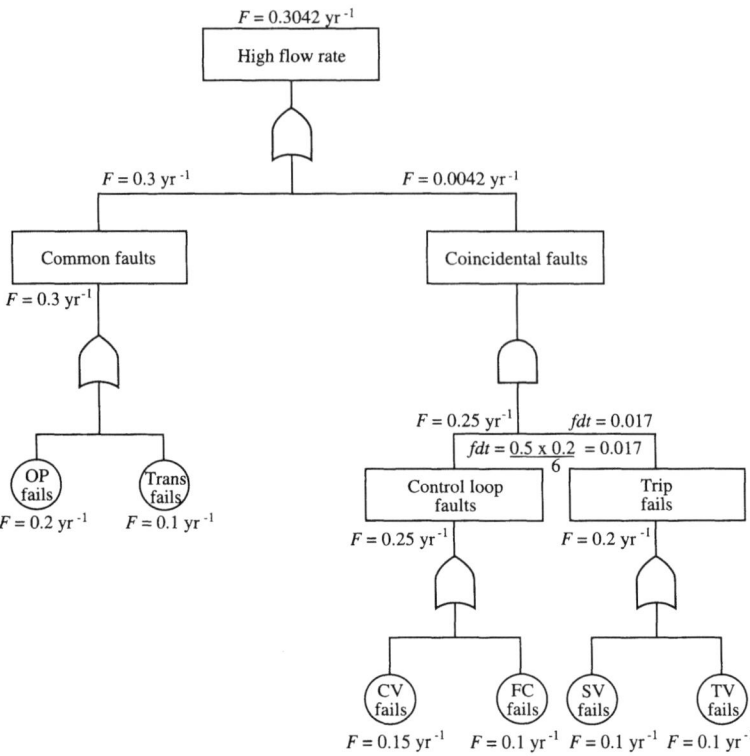

Figure 7.9   Flow control and trip system fault tree.

a process stream by means of the control valve CV. In order to guard against failure of the flow controller and valve (FC and CV), a protective system is provided. This consists of a solenoid valve SV, which takes a signal from the flow transmitter, and a trip valve TV. The failure rates for the components are as follows:

- orifice plate 0.2 yr^{-1}
- flow transmitter 0.1 yr^{-1}
- flow controller 0.1 yr^{-1}
- control valve 0.15 yr^{-1}
- solenoid valve 0.1 yr^{-1}
- trip valve 0.1 yr^{-1}

The trip system is proof-tested every 2 months.

120

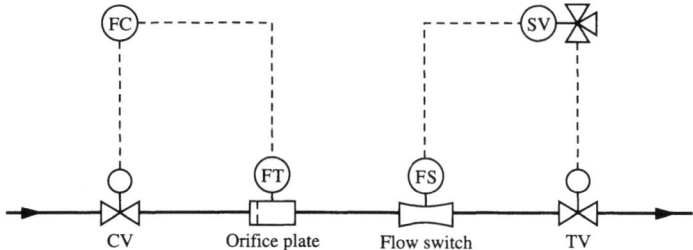

Figure 7.10 Modified flow control and trip system.

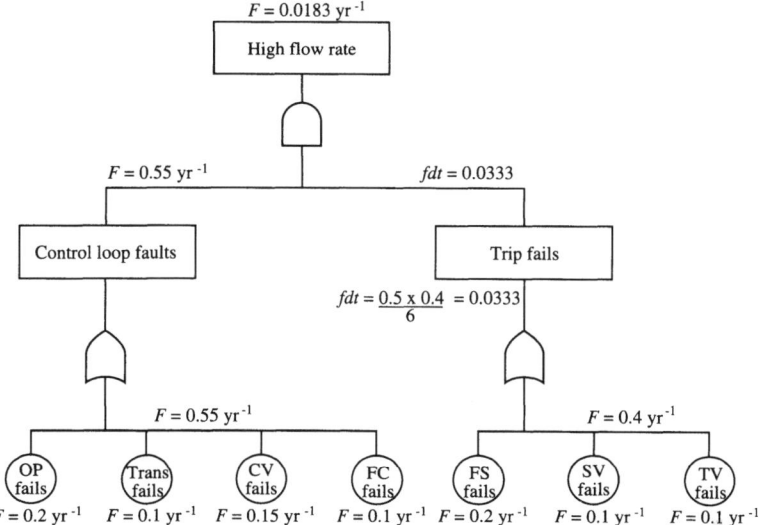

Figure 7.11 Modified flow control and trip system fault tree.

The resulting fault tree is shown in Figure 7.9. This simple example shows how the technique can be used to identify a weak link in a system. The problem clearly rests with the common elements of the system — the orifice plate and the flow transmitter. Figure 7.10 shows an alternative design using an independent flow sensing device to provide the signal to the solenoid valve. The associated fault tree is shown in Figure 7.11.

The resultant top event frequency is reduced from 0.3042 yr^{-1} to 0.0183 yr^{-1}.

EXAMPLE 5

Consider a pair of LPG storage tanks as shown in Figure 7.12. In the event of failure of the level control system the inflow is assumed always to exceed the outflow. The level in the tank will thus rise until it is relieved to the flare header. The flare header has not been designed to take cold LPG and there is a danger that it could block due to the presence of water. There is also a possibility that it will fail due to brittle fracture, thus releasing LPG to the environment. The tanks have level indicators which may alert the operator to a rise in levels, but the operator will not normally be keeping an eye on the indicators. Hence they are not included in the analysis. An independent high level alarm is provided to warn the operator and finally a trip is set to operate from a separate high high level switch.

Acceptance criteria

Assume a basic requirement that the risk of death to employees should not exceed 3×10^{-5} per year. The target hazard rate can then be derived using the following data:

- probability that water will be present in header   100%
- probability that line will rupture   40%
- probability that vapour cloud will ignite   25%
- probability that an employee will be in the area   100%
- probability that a fatality would result from fire/explosion   20%

Hence the maximum allowable frequency for overfilling the tanks is:

$$3 \times 10^{-5} \times \frac{100}{100} \times \frac{100}{40} \times \frac{100}{25} \times \frac{100}{100} \times \frac{100}{20} = 0.0015 \text{ occ yr}^{-1}$$

This figure will be taken as the target for the top event.

Development of logic tree

The top event is 'LPG enters header' and this happens if either of the two tanks overflows. Hence the two tanks will be connected by an OR gate and only one of the tanks will then be followed down the tree. The tank will overflow if the level rises past the high alarm setting to the high high switch and the trip fails, hence an AND gate. The trip is a protective system so it is subject to a regular proof test. The level will reach the high high if a fault develops and either the high level alarm fails or the operator either does not respond or is unable to respond. The level will rise if the control system fails and the operator fails to notice or respond.

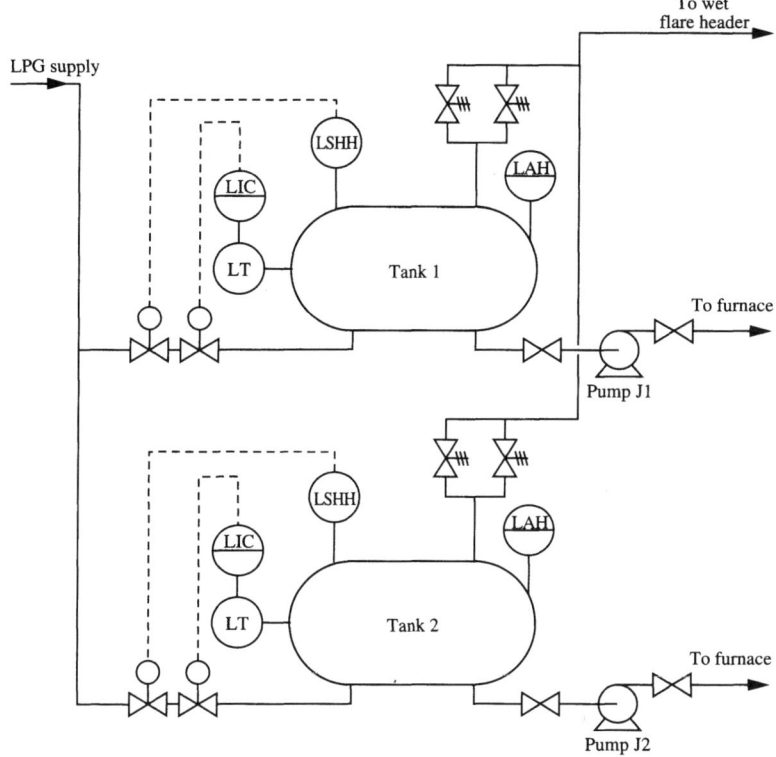

Figure 7.12   LPG storage system.

Calculation of frequencies and probabilities

Control system failure:

- level transmitter or impulse line faults   0.15 occ yr^{-1}
- level controller faults   0.15 occ yr^{-1}
- level control misdirected on manual   0.15 occ yr^{-1}
- control valve faults   0.10 occ yr^{-1}

TOTAL   0.55 occ yr^{-1}

Alarm failure:

- transmitter or impulse line faults   0.15 occ yr^{-1}
- alarm system   0.12 occ yr^{-1}

TOTAL   0.27 occ yr^{-1}

Assume a three month proof test interval, and assume that the trip is disarmed for 30 minutes during testing and the probability of inadvertent isolation after testing is 0.005.

- $fdt = 1/2 \times 0.27 \times 3/12 = 0.03375$
- disarm time $= 30/60 \times 12/3 \times 1/8760 = 0.000228$
- inadvertent isolation $= 0.005$
- total $fdt = 0.03375 + 0.000228 + 0.005 = 0.039$
  $= $ probability of alarm failure (PAF)

Also assume a 5% probability of operator failure (POF). Therefore total probability of 'no remedial action' is given by:

$$PAF + (1 - PAF)POF = 0.039 + (1 - 0.039)0.05 = 0.039 + 0.048 = 0.087$$

Trip failure:

- high level switch fault 0.15 occ yr^{-1}
- solenoid valve fault 0.10 occ yr^{-1}
- trip valve (fail to close) 0.08 occ yr^{-1}
- blocked impulse lines 0.01 occ yr^{-1}
- relay and contact faults 0.01 occ yr^{-1}

TOTAL 0.35 occ yr^{-1}

Again assume a three month test interval and assume that the trip is disarmed for 30 minutes during testing and the probability of inadvertent isolation after testing is 0.005.

- $fdt = 1/2 \times 0.35 \times 3/12 = 0.04375$
- disarm time $= 30/60 \times 12/3 \times 1/8760 = 0.000228$
- inadvertent isolation $= 0.005$
- total $fdt = 0.04375 + 0.000228 + 0.005 = 0.049$

Fault tree

The above figures are put into the fault tree (see Figure 7.13) and combined to give an overfill frequency for a single tank of 0.00234 occ yr^{-1}. As either tank could cause the top event, the total frequency is 0.00468 occ yr^{-1} which is outside the target figure of 0.0015 occ yr^{-1}. Thus further action is required.

It is therefore decided that a duplicate trip system should be provided, identical in all respects to the original system. The modification is shown in Figure 7.14 on page 126.

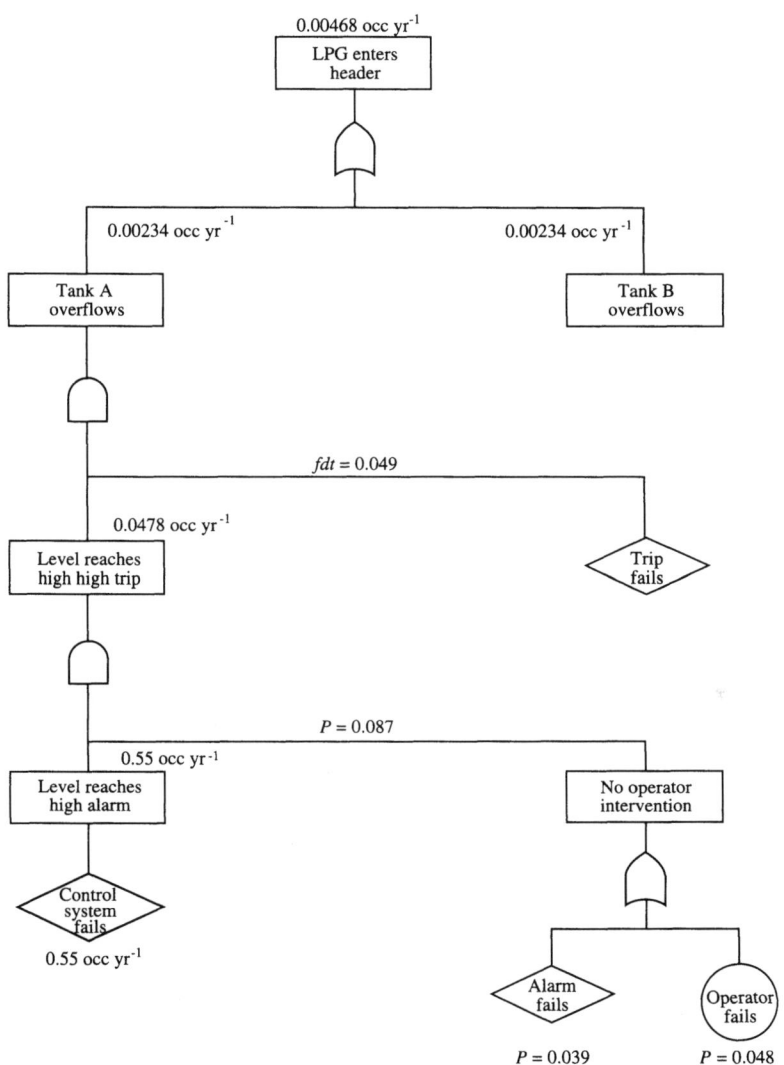

Figure 7.13   LPG storage fault tree.

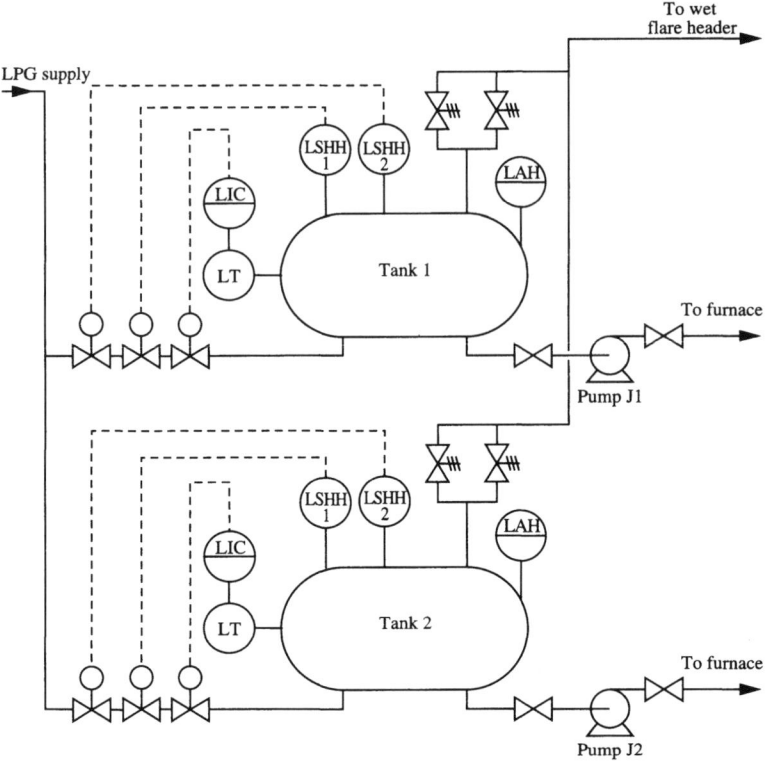

Figure 7.14 LPG storage (twin trips).

Modifications

Allowing for disarm time and test errors, the fractional dead time (*fdt*) for a combined two channel system is given by:

$$fdt_{(1 \text{ out of } 2)} \times (T - 2t)/T + fdt_{(1 \text{ out of } 1)} \times 2t/T$$

where T is the proof test interval = 3/12 = 0.25 yr, and t is the disarm time = $0.5/8760 = 5.71 \times 10^{-5}$.

$fdt_{(1 \text{ out of } 1)}$ = random failure + test errors
= $1/2 \times 0.35 \times 3/12 + 0.005 = 0.04875$

$fdt_{(1 \text{ out of } 2)}$ = coincidence of random faults on both channels +
coincidence of random faults on A and test error on B +
coincidence of random faults on B and test error on A
= $4/3 \times 0.04375^2 + 0.005 \times 0.04375 + 0.005 \times 0.04375$
= 0.00299

126

$$fdt_{(combined)} = 0.00299 \times (0.25 - 5.71 \times 10^{-5})/0.25 +$$
$$0.04875 \times 2 \times 5.71 \times 10^{-5}/0.25$$
$$= 0.00301$$

Allowing for dependent failure ($\beta = 0.2$):

Total probability of failure $= 0.00301 + (0.2 \times 0.04375)$
$$= 0.0118$$

Therefore the frequency for discharge of LPG into the wet flare header becomes:

$$0.0118 \times 0.0478 \times 2 = 0.00113$$

The resulting fault tree is shown in Figure 7.15 on page 128.

This assessment shows that by fitting a '1 out of 2' protection system, the overfill frequency is just within the specified value of 0.0015. The effect of allowing for common mode failure is clearly indicated. The alternative of reducing the proof test interval of a single channel system to, say, one month is not generally acceptable. Had the result still been out of specification, some diverse form of protection would have been required. The results show the root cause of the problem to be the high unreliability of the level control loop and attention is required in this area.

Combination of frequencies
In the above exercise it was assumed that the overflow of either tank would give rise to a hazard. If, however, the flare header was so sized that it required both tanks to overflow at the same time to give rise to a hazard, then it would be necessary to have a knowledge of the likely durations of the overflow as well. If a regular inspection is made of the LPG storage area the person making the inspection is certain to discover that the relief valve has lifted because of the noise it will make. If this inspection takes place twice per shift, say every four hours, and it is possible to stop the feed to the tanks immediately by a means that is 100% reliable, then the duration of relief cannot exceed four hours (4.6×10^{-4} yr). The top end of the fault tree (based on a single protective trip) will now be as shown in Figure 7.16 on page 129. The frequency of the top event is now given by:

$$F_{AB} = F_A.F_B (D_A + D_B)$$
$$= 0.00234 \times 0.00234 \times (0.00046 + 0.00046) = 5 \times 10^{-9} \text{ per year}$$

Thus in this case the company criteria would be satisfied without any additional equipment.

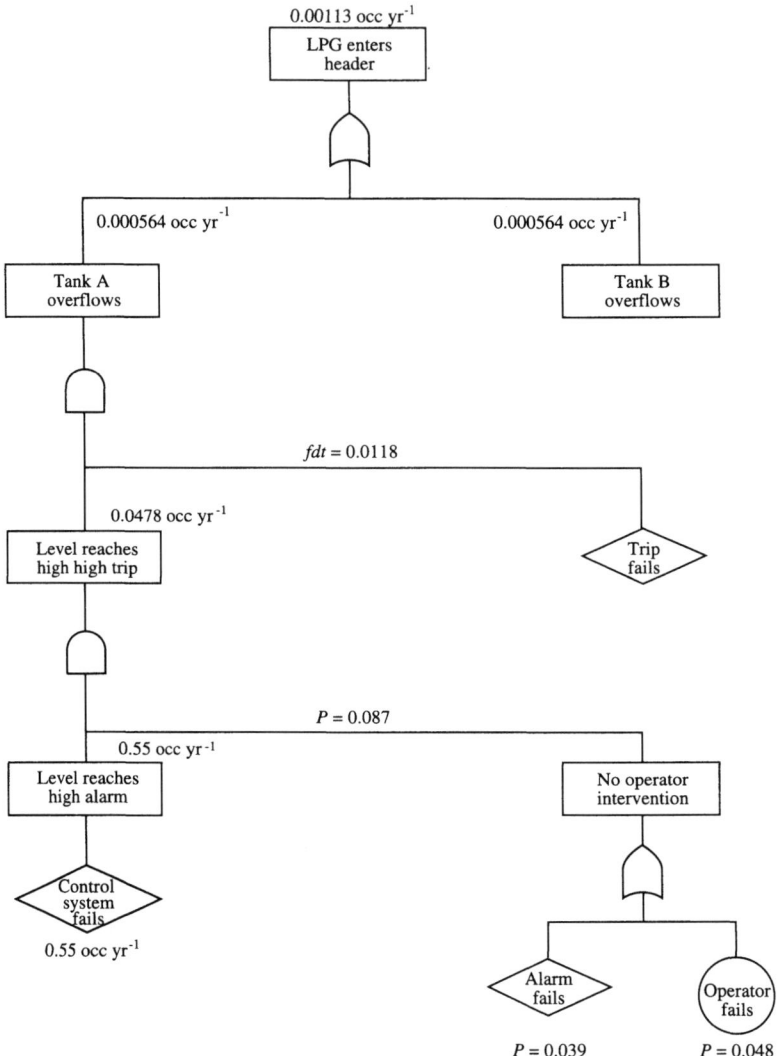

Figure 7.15   LPG storage fault tree (twin trips).

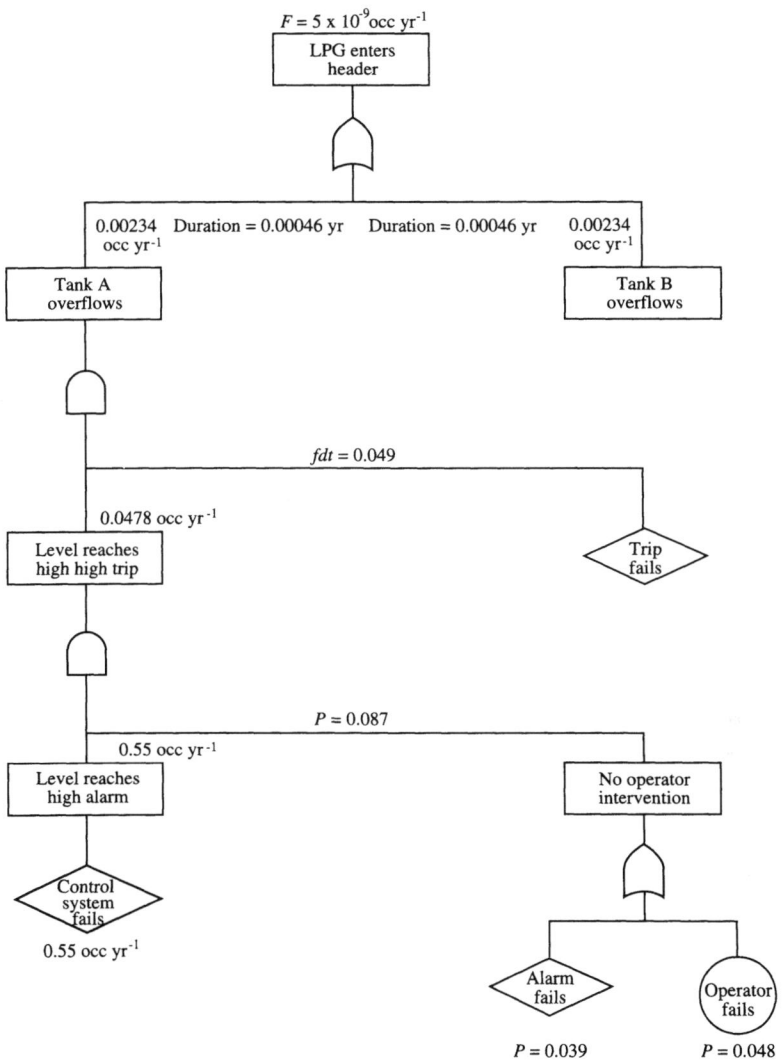

Figure 7.16 LPG storage fault tree for two vessel overflow.

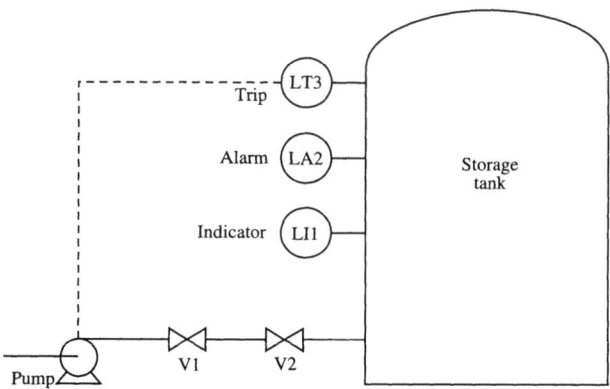

Figure 7.17   Liquid storage system.

7.6   USE OF BOOLEAN ALGEBRA AND CUT SETS

The analysis of complex fault trees is greatly accelerated by the use of the 'cut set' concept and Boolean algebra, as discussed in Chapter 6.

EXAMPLE 6

A liquid storage tank (Figure 7.17) is filled by pump P1. It has a level indicator LI1, a level alarm LA2, and a trip LT3 at successively higher levels. It has two independent shut-off valves V1 and V2, both of which are operator actuated. LI1 is simply an indicator, LA2 has an audible alarm and LT3 automatically trips the pump.

The normal procedure is that the operator observes LI1 and closes V1 when the level reaches L1. Should V1 fail the operator can use V2. The two valves can be considered to be of different design so the question of dependent failure does not arise. If the operator fails to take action an alarm will sound at LA2 and again the operator can close V1 or V2. The ultimate protection is provided by LT3 tripping P1.

The resulting fault tree is shown in Figure 7.18. The minimal cut sets are:

C·D·E
C·K·E
H·D·E
H·K·E
A·B·E

The technique is particularly useful when there are common base events. Thus the failures of V1 and V2 (A and B) appear in two sub-trees but the minimal cut set method ensures that they are combined in the correct way. The same base event cannot appear more than once in a minimal cut set.

The values for each minimal cut set are calculated and the top event frequency found by summation. A further advantage of the technique is that the relative importance of the various minimal cut sets can be found. The concept lends itself to computerization and the cut sets can be listed in order of importance.

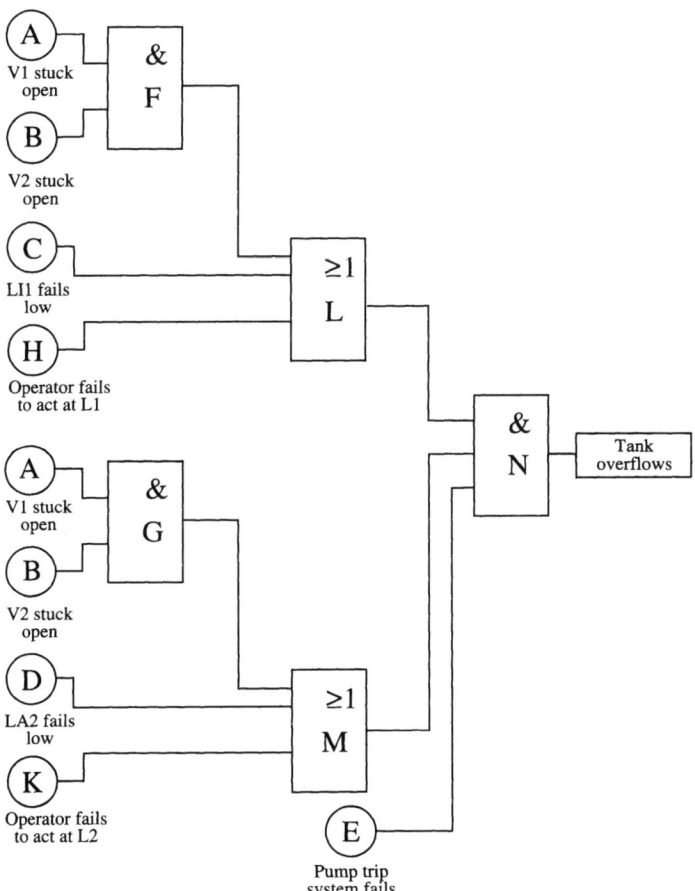

Figure 7.18 Tank overfill fault tree.

Evaluation

The probability of failure for the various events is given below:

A	Valve V1 stuck open	1×10^{-2}
B	Valve V2 stuck open	1×10^{-2}
C	Level indicator LI1 fails low	1×10^{-2}
D	Level alarm LA2 fails	5×10^{-4}
E	Pump trip fails	5×10^{-3}
H	Operator fails to respond to LI	3×10^{-2}
K	Operator fails to respond to LA	1×10^{-2}

$$
\begin{aligned}
C{\cdot}D{\cdot}E &= 10^{-2} \times 5{\times}10^{-4} \times 5{\times}10^{-3} &&= 2.5 \times 10^{-8} \\
C{\cdot}K{\cdot}E &= 10^{-2} \times 10^{-2} \times 5{\times}10^{-3} &&= 5.0 \times 10^{-7} \\
H{\cdot}D{\cdot}E &= 3{\times}10^{-2} \times 5{\times}10^{-4} \times 5{\times}10^{-3} &&= 7.5 \times 10^{-8} \\
H{\cdot}K{\cdot}E &= 3{\times}10^{-2} \times 10^{-2} \times 5{\times}10^{-3} &&= 1.5 \times 10^{-6} \\
A{\cdot}B{\cdot}E &= 10^{-2} \times 10^{-2} \times 5{\times}10^{-3} &&= 5.0 \times 10^{-7} \\
\text{TOTAL} & &&= 2.6 \times 10^{-6}
\end{aligned}
$$

This information clearly shows that the probability of overflow ultimately depends on the pump trip (E) and that operator error (H and K) is also a significant cause of overflow.

It can also be seen that a simple calculation based on the tree as drawn would give a false result because A and B appear in two sub-trees. The tree could either be redrawn or the problem solved by Boolean algebra. The solution using Boolean algebra is as follows:

$F = A{\cdot}B$

$G = A{\cdot}B$

$L = A{\cdot}B+C+H$

$M = A{\cdot}B+D+K$

$N = (A{\cdot}B+C+H)(A{\cdot}B+D+K)E$

$$
\begin{aligned}
N &= A{\cdot}B{\cdot}E + A{\cdot}B{\cdot}C{\cdot}E + A{\cdot}B{\cdot}H{\cdot}E + A{\cdot}B{\cdot}D{\cdot}E + C{\cdot}D{\cdot}E + H{\cdot}D{\cdot}E + A{\cdot}B{\cdot}K{\cdot}E \\
& \quad + C{\cdot}K{\cdot}E + H{\cdot}K{\cdot}E \\
&= A{\cdot}B{\cdot}E + A{\cdot}B{\cdot}H{\cdot}E + A{\cdot}B{\cdot}D{\cdot}E + C{\cdot}D{\cdot}E + H{\cdot}D{\cdot}E + A{\cdot}B{\cdot}K{\cdot}E \\
& \quad + C{\cdot}K{\cdot}E + H{\cdot}K{\cdot}E \\
&= A{\cdot}B{\cdot}E + A{\cdot}B{\cdot}D{\cdot}E + C{\cdot}D{\cdot}E + H{\cdot}D{\cdot}E + A{\cdot}B{\cdot}K{\cdot}E + C{\cdot}K{\cdot}E + H{\cdot}K{\cdot}E \\
&= A{\cdot}B{\cdot}E + A{\cdot}B{\cdot}K{\cdot}E + C{\cdot}D{\cdot}E + H{\cdot}D{\cdot}E + C{\cdot}K{\cdot}E + H{\cdot}K{\cdot}E \\
&= A{\cdot}B{\cdot}E + C{\cdot}D{\cdot}E + H{\cdot}D{\cdot}E + C{\cdot}K{\cdot}E + H{\cdot}K{\cdot}E
\end{aligned}
$$

Note that these are the minimal cut sets derived by inspection.

7.7   TRUNCATION

In order to reduce the amount of work, it may be possible to eliminate many cut sets at a very early stage because they will have little significance in the final result. This must be a matter of judgement but certain rules may be applied. In Example 6 it can be seen that two of the cut sets — those containing event D — have little bearing on the final result. Thus it could be decided that any cut set containing this event can either be taken as negligible or given some arbitrary value, say 10^{-8}. One common rule is to treat as insignificant any cut set with a frequency of less than one thousandth of the value of the highest cut set.

7.8   USE OF INFORMATION

The fault tree will show whether or not the frequency of the top event is within the required criteria. It will also show where changes can best be made to ensure that the criteria are met — that is, it identifies the weak links. So in the example in Figure 7.17 (page 130), if the probability of overflow were too high, the obvious step would be to automate the operator response to the alarm thus changing the highest cut set H.K.E. It can also show where some relaxation may be possible if the required standards are exceeded.

7.9   QUANTIFICATION OF EVENT TREES

The technique of Event Tree Analysis can be used as an alternative calculation of the target frequency in the LPG fault tree in Example 4. Figure 7.19 shows the event tree following the entry of LPG into the header.

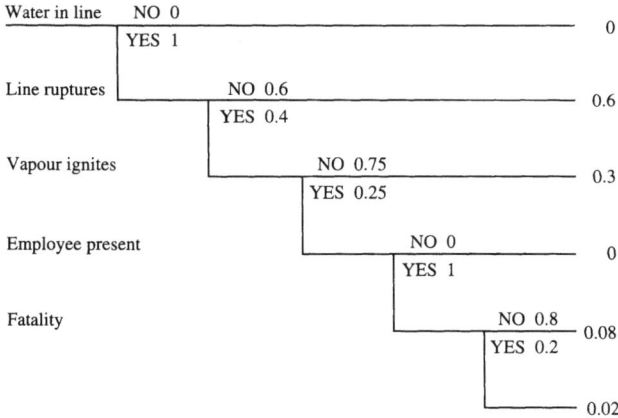

Figure 7.19 LPG storage event tree.

The event tree shows the probability of a fatality to be 0.02. The top event frequency of the fault tree in the unmodified condition is 0.00468, giving a fatal accident frequency of 9.36×10^{-5} which is greater than the company criteria of 3×10^{-5}. The use of an event tree is preferable to a simple calculation because it shows the logic more clearly and enables sensitivity calculations to be performed.

7.10   CONCLUSIONS

In order to ensure a reliable result from a Fault Tree or Event Tree Analysis, give due weight to the following points:

- define the basis for the analysis;
- test the data used for the analysis;
- include all possible basic events capable of leading to the incident;
- check the logic used in combining the various basic events;
- consider the effects of operator error;
- include threats from external sources;
- check for common basic events and other types of dependent failure;
- compare the results with known case histories and check for general reasonableness;
- carry out a sensitivity analysis in any areas of doubtful logic or data;
- present the results in a form suitable for peer group checking.

REFERENCES IN CHAPTER 7

1.   Green, A.E. and Bourne, A.J., 1972, *Reliability Technology* (Wiley Interscience, London, UK).

2.   Lees, F.P., 1980, *Loss Prevention in the Process Industries* (Butterworths, London, UK).

3.   Davidson, J. (ed), 1994, *Reliability of Mechanical Systems*, 2nd edition (IMechE, London, UK).

4.   Flemming, K.N., 1974, *Reliability Model for Common Mode Failures in Redundant Safety Systems, General Atomic Report GA A13284.*

8.   CONSEQUENCE MODELLING

An untoward incident on a process plant may result in the release of a toxic or flammable substance or an uncontrollable release of energy. The release of a flammable substance may in turn result in either an explosion, a fire or both. Injury and damage can be caused by blast, radiated heat and/or the entry of toxic substances into the body. The question of contamination of land and/or food must also be considered. To complete a QRA it is necessary to estimate the consequences of the incident on people and on the surrounding area.

In order to assess the consequences, several factors must be considered, including:

- concentrations of flammable or toxic substance;
- intensity of radiation and/or energy release;
- physical and/or physiological effects;
- local population data.

8.1   GAS DISPERSION

In order to estimate the concentration of dangerous substances it is necessary to examine the ways in which releases of gases and vapours disperse in the environment. With this data it is possible to construct a contour map on which limits of flammability and/or toxicity concentrations can be superimposed. To assess the extent of the dispersion, it is important to know the nature of the release, the height of release above datum, the quantity released, the weather conditions and the local topography.

RELEASE SOURCES

Materials may be released following an incident either as a result of the operation of relief systems — for example, safety valves, blowdown and so on — or as a result of loss of containment — for example, vessel or line rupture. Releases from safety valves and blowdown systems tend to be gaseous, usually entering the atmosphere as jets or plumes at some distance above ground level. Releases arising from loss of containment, however, are frequently liquid or two-phase mixtures entering the atmosphere at or near ground level.

RELEASE RATES

Gas release rates depend on the size of the aperture and the pressure differential. The standard laws of compressible flow apply and many accidental or emergency releases are choked — that is, the velocity remains constant until the pressure differential falls below the critical value. The flow is usually isentropic and adiabatic.

Liquid release rates are more complex, particularly as many liquids are stored under pressure and some may flash on release. This can result in a mixture of vapour and liquid droplets, with the liquid slowly vaporizing in the atmosphere. The basic laws of fluid mechanics apply although tests have shown that at high release rates the effect of flashing may reduce the total emission. If the liquid is below its boiling point, a pool forms and vapour evaporates from the surface of the pool. The estimation of a liquid release is usually based on empirical calculations resulting from test-work. The rate of release can be calculated using one of the following equations:

Gases and vapours

(1) Operating pressure < 2 bar a (assuming the leak is to atmosphere)

$$W = C_d A \{2\rho(P_1 - P_2)\}^{0.5}$$

where:

W = mass flow rate, kg s^{-1};

C_d = coefficient of discharge (usually 0.8);

A = cross-sectional area of orifice, m^2;

P_1 = upstream pressure, N m^{-2} abs;

P_2 = downstream pressure, N m^{-2} abs (= atmospheric for leaks to atmosphere);

ρ = density of fluid, kg m^{-3}.

(2) Operating pressure ≥ 2 bar a

$$W = C_d A P_1 \left[\frac{\gamma M}{R T_1 Z} \left(\frac{2}{\gamma + 1} \right)^{\frac{\gamma + 1}{\gamma - 1}} \right]^{0.5}$$

where:

γ = ratio of specific heats;

R = gas constant = 8314 J kg^{-1} mol^{-1} K^{-1};

M = molecular weight;

T_1 = process temperature, K;

Z = compressibility (= 1 for ideal gas);

other variables as defined above.

Liquids at temperatures below boiling point

$$W = C_d A \{2\rho(P_1 - P_2)\}^{0.5} \text{ kg s}^{-1}$$

Variables as defined above.

Liquids at temperatures above boiling point
If the ratio of path length — that is, the distance travelled by the liquid prior to release — to orifice diameter (l/d) is less than 10 then the flow can be considered to be metastable and the equation for liquids below the boiling point can be used. If the l/d ratio is greater than 10 the flow can be considered as flashing and the following equation must be used:

$$W = C_d A \{2\rho_m(P_1 - P_c)\}^{0.5} \text{ kg s}^{-1}$$

where:

$\rho_m =$ density of equilibrium vapour liquid mixture at P_c, kg m^{-3};
$P_1 =$ upstream pressure;
$P_c =$ 0.55 P_s;
$P_s =$ saturation pressure of liquid at operating temperature, N m^{-2} abs.

$$\rho_m = \frac{1}{\dfrac{M_g}{\rho_g} + \dfrac{1 - M_g}{\rho_1}}$$

where:

$M_g =$ mass fraction of vapour in mixture;
$\rho_g =$ vapour density at P_c and T_c;
$\rho_1 =$ liquid density at P_c and T_c.

$$M_g = \frac{(T_1 - T_c)C_{pl}}{L}$$

where:

$T_1 =$ process temperature of liquid, K;
$T_c =$ equilibrium temperature of vapour/liquid mixture in K at P_c;
$C_{pl} =$ average specific heat of liquid between T_1 and T_c, kJ kg^{-1} K^{-1};
$L =$ average latent heat of vaporization between T_1 and T_c, kJ kg^{-1}.

Vaporization of liquid releases
When a hot liquid is released into the atmosphere some of it vaporizes; the percentage vaporized may be calculated as follows:

$$V_v = \frac{100(T_1 - T_2)C_{pl}}{L} \%$$

where:

V_v = mass percentage of liquid vaporized;

T_2 = final temperature, °C (or, if that is not known, the boiling point of the liquid at atmospheric pressure);

other variables as defined above.

For flashing releases the above value is doubled to allow for the effect of spray vaporization.

Vaporization of liquid pools

It is first necessary to calculate the area of the pool. One commonly used equation is:

$$A_p = \frac{Vt}{h_m}$$

where:

V = volumetric flow, $m^3 s^{-1}$;

t = time, s;

h_m is a pool depth factor = 5×10^{-3} m for concrete and 10^{-2} m for gravel.

The vaporization of a pool of liquid depends on a number of factors (see References 1 and 2 for a full treatment). However, taking a conservative case of Pasquill stability F (see Table 8.1 on page 140 for definitions of Pasquill stability) and a wind speed of 2 m s^{-1}, the initial rate of vaporization is given by:

$$G = 3.6 \times 10^{-3} \, \rho_v \frac{P_v}{P_a} \left(\frac{0.5}{D_p}\right)^{0.18}$$

where:

G = rate of vaporization, kg s^{-1} m^{-2} surface area;

ρ_v = vapour density, kg m^{-3};

P_v = vapour pressure of pool, N m^{-2};

P_a = atmospheric pressure, N m^{-2};

D_p = diameter of pool, m.

This equation gives the initial rate of vaporization; the rate falls as the pool chills due to vaporization. In general the rate of vaporization from a leak is much greater than that from the resulting pool, so the pool is often neglected. This is particularly true indoors where there is little movement of air[2].

APERTURE AREA

One problem associated with both gaseous and liquid releases is the estimation of the aperture area. For pipelines two conditions are normally considered — a full bore or guillotine break, and an aperture of area equal to that of one quarter of the diameter (d/4 break). In the case of a guillotine break it is assumed that the fluid exits from both ends of the break. For safety valves, bursting discs and so on, the area and characteristics are known and reasonable estimates can usually be made. For flange leaks the aperture is normally taken as the area formed by the flange thickness and the distance between two bolts. For seals an estimate is made based on shaft diameter and a radial gap (usually 0.125 mm). For vessel rupture, empirical correlations are necessary. Reference 1 gives a good comparison of methods of estimating release rates.

DISPERSION PATTERN

The next problem is to determine what will happen once the gas or vapour is released. The behaviour is very complex depending on the manner of release, height of release and atmospheric and topographical factors. Atmospheric conditions are particularly important and it is essential to understand something of the nature of the atmosphere close to the surface of the earth.

ATMOSPHERIC FACTORS

An atmospheric layer known as the planetary boundary layer extends up to about 1 km from the surface of the earth. Flow patterns in this layer are influenced by drag from the rotation of the earth and are predominately turbulent. The adiabatic vertical temperature gradient (($\mathrm{d}T/\mathrm{d}z)_{ad}$ — approximately 0.98°C per 100 m) can be disturbed leading to movement of air or the trapping of emissions and pollutants. The temperature gradient varies diurnally, the actual variation depending on weather conditions and location. If the sky is clear during the day, the ground is heated by solar radiation and the air adjacent to it will be warmer than the adiabatic gradient. At night the effect is reversed. Figures 8.1(a) and (b) on page 140 show typical profiles at dawn and early afternoon with a clear sky. At coastal locations the fact that the sea temperature varies less than the land temperature complicates the situation.

If the temperature distribution of a column of air matches that of the atmosphere it is said to be neutral. If the temperature of the air column changes less than that of the atmosphere it is said to be stable (Figure 8.1(a)) and if it changes more than that of the atmosphere it is said to be unstable (Figure 8.1(b)). Stable conditions tend to trap emissions of noxious substances whereas unstable conditions assist dispersion.

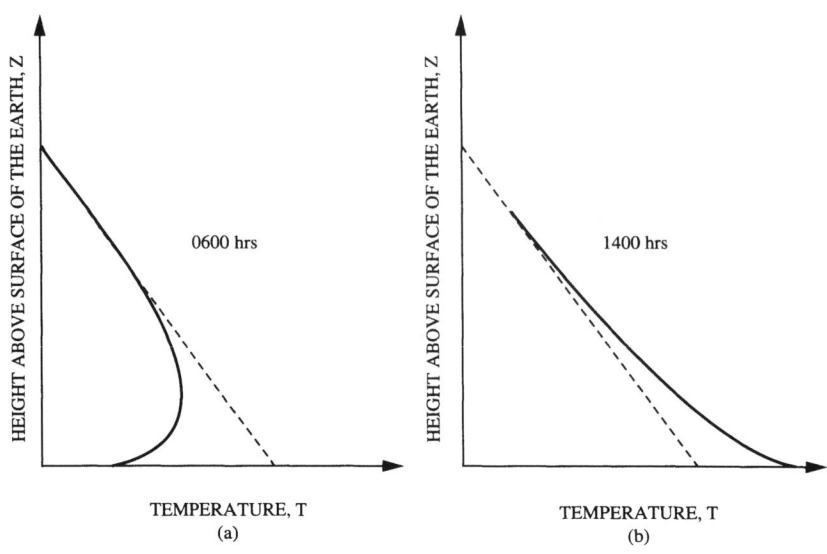

Figure 8.1   (a) Stable conditions. (b) Unstable conditions.
$dT/dz < (dT/dz)_{ad}$: unstable. $dT/dz = (dT/dz)_{ad}$: neutral. $dT/dz > (dT/dz)_{ad}$: stable.

TABLE 8.1
Pasquill stability class[†]

Surface wind speed, m s^{-1}	Daytime insolation			Night conditions	
	Strong	**Moderate**	**Slight**	**Cloudy**	**Clear**
< 2	A	A–B	B		
2–3	A–B	B	C	E	F
3–4	B	B–C	C	D	E
4–6	C	C–D	D	D	D
> 6	C	D	D	D	D

[†] Crown Copyright, reproduced with the permission of the Controller of HMSO

The dispersion pattern is also affected by the wind velocity and the two sets of atmospheric conditions are combined in the Pasquill stability class shown in Table 8.1[3].

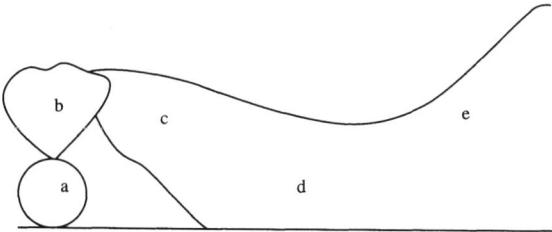

Figure 8.2   Stages in the release and dispersion of a cold or dense gas.
a — source of emission; b — initial momentum; c — internal buoyancy;
d — transition; e — ambient turbulence.

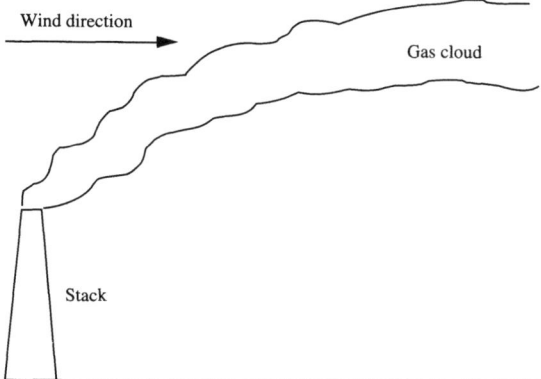

Figure 8.3   Stack release.

DISPERSION PATTERNS

Gases released as plumes or jets have an upward momentum and may have buoyancy due to their temperature. Figure 8.2 shows typical stages in the release and dispersion of a gas. The initial momentum results in some degree of dilution due to entrainment; assuming the gas is heavier than air it tends to fall as the effects of internal buoyancy predominate. As the gas cloud becomes further diluted, the effect of natural atmospheric turbulence takes control and further spread depends upon atmospheric and topographical conditions.

Gases released from stacks follow the trajectory shown in Figure 8.3. A number of equations have been developed to describe that trajectory in terms of efflux velocity, wind velocity, gas properties and so on.

141

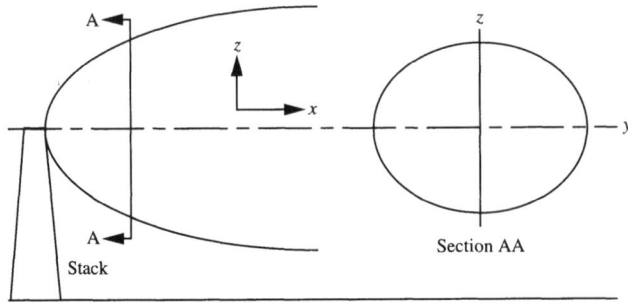

Figure 8.4   Gaussian plume model.

Dense gas plume behaviour depends on the ratio of the potential energy of the plume (due to its excess density) to the turbulent energy of the atmosphere. This factor is quantified by the Richardson Number:

$$Ri \; = \; \frac{g}{T} \frac{dT/dz + \Gamma}{(du/dz)^2}$$

where:

g = acceleration due to gravity, m s^{-2};

T = absolute temperature, K;

u = wind speed, m s^{-1};

z = vertical distance, m;

Γ = dry adiabatic temperature lapse, K m^{-1}.

The motion is turbulent if $Ri < 1$ and laminar if $Ri > 1$. Dense gas effects are likely to predominate if $Ri \gg 1$.

Further away from the stack the initial effects no longer predominate and the behaviour is that of a neutrally buoyant plume. At this point the Gaussian plume model (Figure 8.4) may be used to predict behaviour and hence allow an estimate of concentration patterns. The actual formulae are very complex and include diffusion coefficients based on the Pasquill stability class. Calculations are usually performed using computer models which include factors for topography and climatic conditions.

The equation giving the concentration for a continuous elevated point source is:

$$C_{x,y,z} = \frac{Q}{2\pi\sigma_y\sigma_z u} \exp\left(-\frac{y^2}{2\sigma_y^2}\right)\left[\exp\left(-\frac{(z-H)^2}{2\sigma_z^2}\right) + \exp\left(-\frac{(z+H)^2}{2\sigma_z^2}\right)\right]$$

where:

142

Q = continuous mass release rate, kg s^{-1};

H = stack height, m;

u = wind velocity, m s^{-1};

σ_y and σ_z = diffusion coefficients.

The diffusion coefficients depend on the Pasquill stability class and can be obtained from Figure 8.5, taken from Reference 4.

The ground level concentration is given by:

$$C_{x,y} = \frac{Q}{\pi \sigma_y \sigma_z u} \exp\left(-\frac{y^2}{2\sigma_y^2}\right)\left[\exp\left(-\frac{(H^2)}{2\sigma_z^2}\right)\right]$$

The basic equation for continuous release at ground level is:

$$C_{x,y,z} = \frac{Q}{\pi \sigma_y \sigma_z u} \exp\left[-\frac{1}{2}\left(-\frac{y^2}{\sigma_y^2} - \frac{z^2}{\sigma_z^2}\right)\right]$$

For an instantaneous release at ground level (Q^*) the concentration varies with time as well as distance in accordance with the following equation:

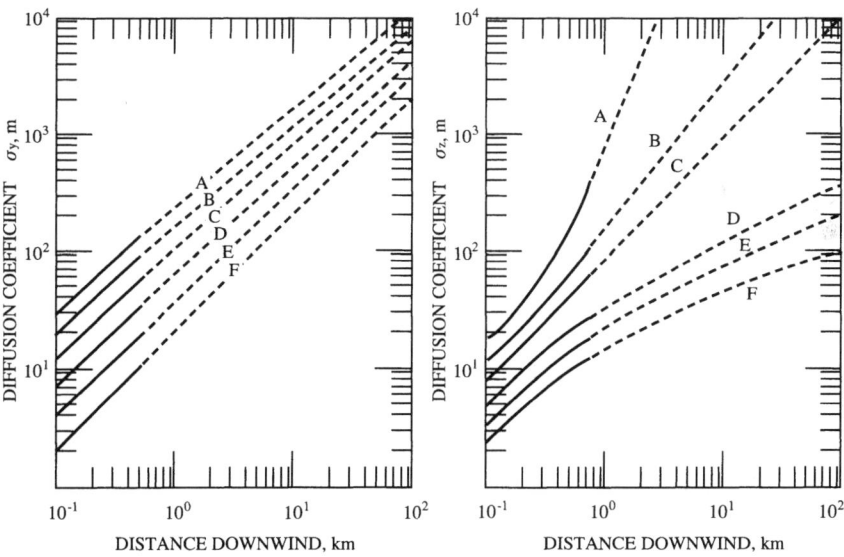

Figure 8.5   Values of diffusion coefficients.

143

$$C_{x,y,z,t} = \frac{2Q*}{(2\pi)^{3/2}\sigma_x\sigma_y\sigma_z} \times \exp\left[-\frac{1}{2}\left(-\frac{(x-ut)^2}{\sigma_x^2} + \frac{y^2}{\sigma_y^2} + \frac{z^2}{\sigma_z^2}\right)\right]$$

where:

$Q*$ = instantaneous release, kg;

t = elapsed time, s.

Releases at or near ground level are strongly influenced by ground conditions and topography. Table 8.2 (taken from Reference 5) gives some values of diffusion coefficients for different topography. The above equations should be used with some caution as they assume that the gas being dispersed has the same density as air. This is obviously not the case for many chemical releases. Use of CFD techniques is now preferred.

Dense gas releases are dominated by the gravity slumping effect. The main factors are shown in Figure 8.6.

It is clear that the effects are extremely complex with both atmospheric and surface conditions playing a part. Once the cloud touches the ground there is heat transfer with the surface which may heat or cool the cloud depending on the temperature difference.

TABLE 8.2
Formulae for diffusion coefficients in terms of distance 'x'[†]

Pasquill type	σ_y, m	σ_z, m
Open country		
A	$0.22x(1+0.0001x)^{-0.5}$	$0.20x$
B	$0.16x(1+0.0001x)^{-0.5}$	$0.12x$
C	$0.11x(1+0.0001x)^{-0.5}$	$0.08x(1+0.0002x)^{-0.5}$
D	$0.08x(1+0.0001x)^{-0.5}$	$0.06x(1+0.0015x)^{-0.5}$
E	$0.06x(1+0.0001x)^{-0.5}$	$0.03x(1+0.0003x)^{-1}$
F	$0.04x(1+0.0001x)^{-0.5}$	$0.016x(1+0.0003x)^{-1}$
Urban conditions		
A–B	$0.32x(1+0.0004x)^{-0.5}$	$0.24x(1+0.001x)^{0.5}$
C	$0.22x(1+0.0004x)^{-0.5}$	$0.20x$
D	$0.16x(1+0.0004x)^{-0.5}$	$0.14x(1+0.0003x)^{-0.5}$
E–F	$0.11x(1+0.0004x)^{-0.5}$	$0.08x(1+0.00015x)^{-0.5}$

[†] Reproduced from Briggs, G.A., 1973, *Diffusion Estimation for Small Emissions,* *ATDL Contribution File No 79.*

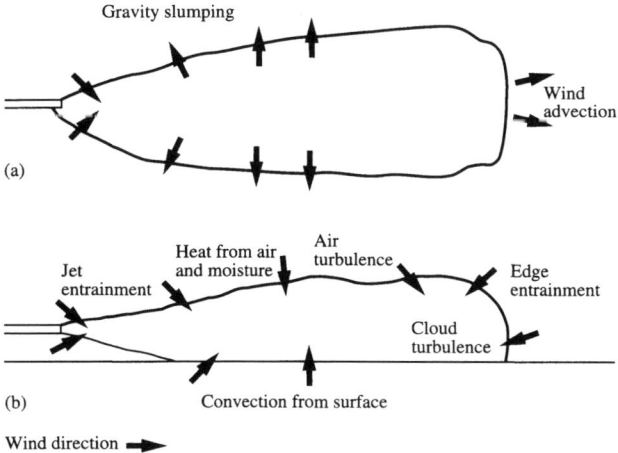

Figure 8.6   Dense gas dispersion: (a) view from above; (b) view from side.

8.2   TOXICITY

Using the models now available, it is possible to determine the variation in concentration with distance from the source as a function of time. Medical statistics show that for most harmful substances:

$$\text{immediate effect} \propto (\text{concentration})^n \times \text{time}$$

The index n is usually greater than unity; a value of 2.75 is taken for chlorine. Occupational exposure limits (OELs)[6] can be taken as a guide to show whether or not the effects of an incident are hazardous.

Information on long-term effects is much less clear, particularly for complex organic chemicals and suspected carcinogens. Medical statistics are still being gathered and are often disputed, particularly where claims for compensation are involved. Work based on animal experiments is often quoted but its applicability to humans must be considered very carefully.

POPULATION DATA

Further factors which influence the consequences of the release of dangerous substances are the density and structure of the population affected by the release and whether the people are likely to be indoors or out of doors. Medical information confirms that the very young and the very old are much more susceptible to harmful effects than are most other sectors of the population. Thus the release

145

of a noxious substance close to a school or old peoples' home will have a higher casualty potential. Large outdoor concentrations of people are more likely to be affected than those indoors because some degree of protection can be obtained by closing doors and windows. Mobility is another factor; concentrations of people who cannot be readily evacuated — such as those in prisons and hospitals — are more likely to be affected by a toxic release than the population in general. Finally, in this part of the calculation, consideration must also be given to the proportion of time for which a particular area is likely to be occupied. Thus a sports stadium which is occupied only for short periods poses less risk than a housing development which is occupied continuously.

8.3   EXPLOSIONS AND FIRES

INTRODUCTION
The consequences of an escape of flammable material may be either an explosion or a fire or, possibly, both. It is thus important to be able to predict the likelihood of such explosions and/or fires and the extent of the resultant damage. Explosions can also be caused by dusts as well as by flammable gases, vapours and solid explosives (usually known as condensed explosives). The questions of prevention and mitigation have already been discussed in Chapter 2.

DEFINITIONS

Explosion
The uncontrolled release of energy from a flame front propagating through a flammable medium, and characterized by the generation of heat, light and pressure.

Unconfined explosion
Ignition of a flammable mixture in the open atmosphere.

Confined explosion
Ignition of a flammable mixture in a sealed container — for example, a building or section of plant.

Vented explosion
Partially confined explosion where there exists a vent through which combustion gases can escape.

Flammability limits

A mixture of fuel and oxidant will only burn if the fuel concentration lies within a certain range. The top and bottom of the range are known as the upper and lower flammability limits. Flammability limits widen as temperature increases.

Laminar burning velocity, S_u

The velocity with which a one-dimensional plane combustion wave propagates through an infinite flammable mixture relative to that mixture. S_u is a fundamental property of any flammable mixture, and increases with decreasing pressure and increasing temperature.

Flame speed, S_f

The rate of propagation of a flame front with respect to a fixed observer. Usually S_f differs from S_u owing to expansion/turbulence effects ahead of the flame front. Note $S_f = 0$ for a stationary flame.

Adiabatic flame temperature

The temperature that would be attained by a flame if there were no losses by radiation or convection to the containment.

Expansion factor, E

A measure of the increase in volume resulting from combustion.

$$E = \frac{\rho_r}{\rho_p} = \left(\frac{T_p}{T_r}\right)\left(\frac{N_p}{N_r}\right)$$

where:

ρ = density of mixture, kg m^{-3};

T = temperature, K;

N = number of moles of mixture;

subscript r = reactants;

subscript p = products.

Auto-ignition temperature

The temperature at or above which a flammable mixture will ignite spontaneously.

Minimum ignition energy

The minimum energy that is required to ignite a flammable mixture.

147

Deflagration
An explosion where the flame speed is less than the speed of the pressure wave (usually a few metres per second), and wherein propagation is limited by heat and mass transfer processes.

Detonation
An explosion where the flame front is coupled with a high pressure shock wave, thereby causing very high flame speeds in excess of the speed of sound (usually 1.5 to 3 km s^{-1}).

EXPLOSION PRESSURE
An explosion generates a rapid rise in pressure and the wave propagates causing damage in its path. It is then followed by a negative pressure wave which causes further damage before the pressure returns to atmospheric. Thus the damage depends on the maximum pressure reached and the velocity of propagation. It will obviously also depend on the physical characteristics of the environment concerned. Figure 8.7 shows a typical pressure time curve. There is a further peak pressure of extremely short time duration (the von Neumann spike) but the duration is so short that it does not have any engineering significance. In practice the impulse — that is, the pressure time area beneath the curve — is a better measure of the severity of the explosion.

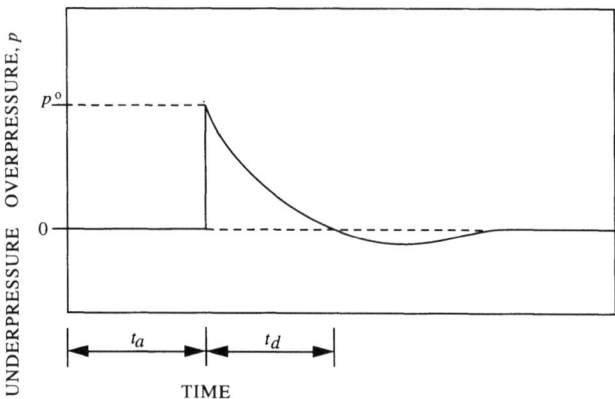

Figure 8.7   Variation in explosion overpressure with time at a fixed point.

TABLE 8.3
Flammability data

Substance	LEL, %	UEL, %	Maximum S_u, m s^{-1}	Maximum S_f, m s^{-1}	Auto-ignition Temp, °C	Minimum ignition energy, mJ
Hydrogen	4.0	75.0	3.2	22.1	400	0.01
Methane	5.0	15.0	0.37	2.8	540	0.26
Ethane	3.0	12.4	0.44	3.4	515	0.24
Propane	2.1	9.5	0.42	3.3	450	0.25
n-butane	1.8	8.4	0.42	3.3	405	0.26
Pentane	1.4	7.8	0.42	3.4	260	0.22
Hexane	1.2	7.4	0.42	3.4	225	0.23
Acetylene	2.5	100	1.7	14.8	305	0.02
Ethylene	2.7	36	0.83	6.5	490	0.12
Propylene	2.4	11.0	0.48	3.8	460	0.28
Benzene	1.3	7.9	0.62	5.0	560	0.22
c-hexane	1.3	8.0	0.52	4.2	245	0.24

A deflagration can produce pressure rises of about 8 to 1; much higher rises are produced by a detonation. Of equal importance is the rate of pressure rise, dp/dt. This depends on the way in which the explosion is contained and on the characteristics of the mixture.

GAS AND VAPOUR EXPLOSIONS

Gas dispersion data together with information on upper and lower explosion limits can be used to calculate the likelihood of an explosion. A flammable mixture only explodes if its concentration is between these two limits and it only ignites without an external source of ignition if its temperature is above the auto-ignition temperature. It is, however, customary in safety analysis to assume the presence of a source of ignition unless there is evidence to the contrary. In order for a mixture to ignite below its auto-ignition temperature the ignition source must have a minimum energy level.

Table 8.3 gives information of upper and lower explosion limits and other data useful in assessing the likelihood and effects of an explosion.

TABLE 8.4
K values for gases

Gas	K_v, bar m s^{-1}
Methane	55
Propane	75
İHydrogen	550

For mixtures the limits can be calculated using the following equation:

$$FL_{mix} = \frac{1}{\sum\limits_{i=1}^{n} \dfrac{y_i}{FL_i}}$$

In general the flammability range increases with temperature but is not significantly affected by pressure.

The rate of pressure rise in an isothermal system follows a cube law and is given by:

$$P = \frac{\text{const} \times S_u^3 \, t^3 \, P_{max}}{V}$$

where:

S_u = laminar burning velocity, m s^{-1};
P_{max} = maximum pressure reached when combustion is complete, N m^{-2};
V = volume of the containment, m^3.

In practice a more useful equation is:

$$\frac{dp}{dt} \, V^{0.33} = K$$

K, sometimes called the deflagration index (K_v for vapours and gases), can be determined experimentally. Typical values are given in Table 8.4.

DUST EXPLOSIONS
The analyses of dust explosions are rather more complex than those for gas/vapour explosions but are just as important because dust explosions are a considerable hazard in the process industries. A very wide range of dusts are

150

TABLE 8.5
Explosion data on dusts[†]

Dust	Cloud ignition energy, mJ	Ignition cloud, °C	Temp layer, °C	Maximum explosion pressure, bar a	Maximum rate of pressure rise, bar s^{-1} m^{-3}	Minimum explosion conc	Limiting oxygen conc, vol %
Lignite	30	390	180	11.0	151	60	12
Aluminium	15	550	740	13.0	750	60	5
Coal	60	610	170	9.8	114	15	14
Cellulose	80	480	270	11.0	125	30	9
Cornflour	40	380	330	10.3	125	60	9
Wood	40	470	260	10.2	142	60	10
Charcoal	20	530	180	10.0	10	60	
Sugar	30	370	400	9.5	138	60	
Sulphur	15	190	220	7.8	151	30	
Magnesium	80	450	240	18.5	508	30	
Zinc	9600	690	540	7.8	93	250	

[†] Based on data published in Lunn, G.A., 1992, *Dust Explosion Prevention and Protection, Part 1 — Venting* (IChemE, Rugby, UK).

explosible and Table 8.5 gives some data on typical dusts. Remember that individual dusts vary considerably and tests should be carried out on representative samples whenever possible.

A dust explosion will occur if:
- the dust is explosible;
- the size distribution allows propagation of flame;
- the atmosphere in which the dust is dispersed contains sufficient oxidant;
- the dust cloud concentration lies within the explosible range;
- the dust cloud is in contact with an ignition source of sufficient energy.

Dust explosions follow the same pressure rise equations as vapour clouds. Typical values of the deflagration index K for various types of dust are given in Table 8.6 on page 152. Note that dusts are classed (St class) according to their K values.

EXPLOSION YIELD
In order to estimate the effects of an explosion, some way of quantifying the yield must be found. The energy released in an explosion as a blast wave is

TABLE 8.6
K values for dusts[†]

Explosion class	K_{st}, bar m s^{-1}	Characteristic	Example
St 0	0	Non-explosible	Stone dust
St 1	0–200	Explosible	Milk powder
St 2	200–300	Strongly explosible	Wood dust
St 3	> 300	Very strongly explosible	Aluminium dust

[†] Based on data published in Lunn, G.A., 1992, *Dust Explosion Prevention and Protection, Part 1 — Venting* (IChemE, Rugby, UK).

generally only a fraction of the total available (10% to 25% for solid materials and up to 50% for gases confined in a pressure vessel). Estimates of damage can be based on experience, including military test work, or based on semi-empirical calculations. Much of the early work in this area was done by the military authorities to assess the performance of weapons using the standard unit of TNT equivalent. Other substances are compared on the ratio of heat of combustion of the substance to the heat of detonation of TNT (4652 kJ kg^{-1}), the energy ratio. Although this method is disputed by some specialists, no satisfactory alternative has yet been devised.

The overpressure caused depends on the amount and nature of the material involved and the distance to the affected area or target (scaled distance). The first stage is to estimate the 'overpressure' at the target using Figure 8.8 (Reference 7), where the mass is of an equivalent amount of TNT.

The approach for vapour cloud explosions still uses the TNT equivalence concept but the actual calculation of the TNT equivalence allows for the fact that only a very small amount of the available energy is actually released as a blast wave (between 3% and 10%). Table 8.7 gives some typical data. The results are likely to give an overestimate of the overpressure, particularly close to the source. The method is not applicable to pipes because of the complex way in which pressure waves propagate along a pipe.

Note that Figure 8.8 is based upon flat terrain. Upward slopes can cause overpressure enhancement and downward slopes tend to reduce the overpressure. The figure is extrapolated from military data and will tend to overestimate the overpressure as most explosive accidents do not involve actual explosives.

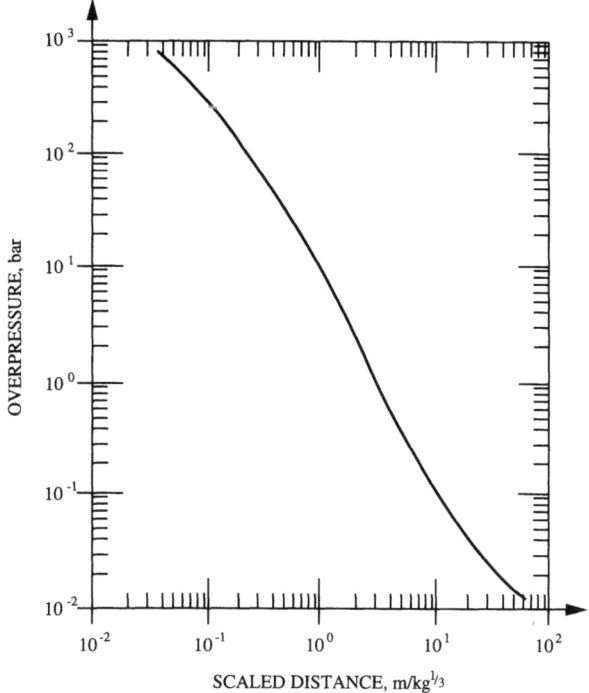

Figure 8.8   Overpressure due to explosions.

TABLE 8.7
TNT equivalence for vapour clouds

Material	Energy ratio	Efficiency factor	TNT equivalence
Hydrocarbons	10	0.04	0.4
Ethylene oxide	6	0.1	0.6
VCM	4.2	0.04	1.16
Acetylene oxide	6.9	0.06	0.4

153

For flashing liquids the TNT equivalence must be based on the amount of vapour actually present in the cloud. It is calculated as follows:

$$W_v = \frac{2W_1 C_{pl}(T_1 - T_{bp})}{L}$$

where:

$W_v =$ weight of vapour, kg;
$W_1 =$ weight of liquid involved, kg;
$C_{pl}=$ specific heat of the liquid at $(T_1 - T_{bp})/2$, kJ kg^{-1} °C^{-1};
$T_1 =$ temperature of the liquid in the plant, °C;
$T_{bp}=$ boiling point of the liquid at atmospheric pressure, °C;
$L =$ latent heat of vaporization at T_{bp}, kJ kg^{-1}.

The factor of 2 is an arbitrary figure to allow for further vaporization of the liquid spray in air.

Example
What would be the overpressure at a distance of 300 m if a 30 tonne tank of propane ruptured and the vapour cloud exploded?

Ambient temperature, T_1 = 20°C
Specific heat, C_p = 2.41 kJ kg^{-1} K^{-1}
Boiling point, T_{bp} = − 42°C
Latent heat, L = 410 kJ kg^{-1}

$$W_v = \frac{2W_1 C_p(T_1 - T_{bp})}{H_v} = \frac{2 \times 30 \times 2.14 \times (20 - [-42])}{410} = 21.87 \text{ tonnes}$$

TNT equivalence (from Table 8.6) = 0.4
Equivalent mass of TNT = 8.75 tonnes
Scaled distance = 300/(8750)$^{0.333}$ = 14.6
Overpressure (from Figure 8.8) = 0.055 bar

DUSTS
The yields of dust explosions are generally much below those of gas or condensed phase explosions, and dust explosions are more likely to be deflagrations rather than detonations. They can still do a substantial amount of damage if steps are not taken to suppress, vent or contain them.

154

CONDENSED PHASE EXPLOSIONS

Condensed phase — for example, solid material explosions — may be either deflagrations or detonations. There is a minimum physical size or charge diameter below which an unconfined detonation is not possible. This is due to the fact that surface energy losses may be sufficient to prevent supersonic reaction velocities. This factor is important in the design of military and commercial explosives; the critical diameter ranges from below 1 mm for nitro-glycerine to over 1 m for some blasting explosives.

DAMAGE CAUSED BY EXPLOSIONS

Once the overpressure is known, Tables 8.8 and 8.9 can be used to estimate the damage likely to be caused by the explosion. These two tables over-simplify the

TABLE 8.8
Damage effects of explosions

Damage	Overpressure, bar
Houses almost demolished (75% destroyed)	0.4
Houses severely damaged (50% destroyed)	0.25
Houses rendered temporarily uninhabitable	0.1
Doors and windows shattered	0.07
50% window breakage	0.025
Ground cleared	2.0
Trunks or large branches of trees broken off	0.8
Utility poles snapped	0.7

TABLE 8.9
Effects of overpressure on people

Overpressure, bar	Casualty probability
< 0.07	0
0.07–0.21	0.1
0.21–0.34	0.25
0.34–0.48	0.7
> 0.48	0.95

problem because information is also needed on the pressure/impulse relationship which depends on the type of explosion and the scaled distance. Human casualties result from being blown over and from collapsing buildings as well as from direct overpressure.

In addition to damage caused by the blast, consideration must also be given to damage caused by missiles generated by the explosion (shrapnel). Exploding pressure vessels usually cause more damage by fragmenting than by blast. Empirical correlations are available to estimate the number of fragments or missiles generated and consideration must also be given to any secondary missiles generated by the blast or impact of primary missiles. Critical plant items are often protected against missile damage in order to prevent the escalation of incidents. For further information see Reference 8.

PHYSICAL EXPLOSIONS
Another form of explosion is the rupture of a vessel or pipe due to excessive pressure not generated by chemical reaction. Such explosions are usually relatively slow and the energy released may be estimated from the Brode equation:

$$E = \frac{(P_i - P_f)\, V}{(\gamma - 1)} \; \text{Joules}$$

where:

P_i = initial pressure, N m^{-2};

P_f = final pressure, N m^{-2};

V = vessel volume, m^3;

γ = ratio of the specific heats, C_p/C_v.

8.4   FIRES

Fire can cause damage by direct impingement or by radiation. Fires can be of a number of types:
- pool — a burning pool of liquid;
- torch — a jet of flame issuing from, say, a ruptured pipe;
- flash — a sudden rapid combustion of a cloud of gas (really an explosion);
- BLEVE — boiling liquid expanding vapour explosion (fireball);
- running — a cascade of burning fuel down a structure, etc.

A key factor in determining the type of fire is the volatility of the liquid. A liquid with low volatility burns as a pool fire or, if released under pressure, as a torch or jet fire. Liquids with higher volatility form significant clouds of vapour resulting in flash fires or fireballs. Sometimes there is a combination as the 'rain out' of liquid droplets from a flashing vapour gives rise to both a flash fire and a pool fire on the ground below.

Diffusion flames can reach temperatures of up to 1000°C and torch flame temperatures can be much higher. Depending on conditions, most metals start to lose their mechanical strength in about 10 minutes if subjected to direct flame impingement. This can lead to rupture of vessels and loss of integrity of structural steelwork.

Calculation of heat flux is rather complex and the value depends on the combustion rate, the calorific value of the fuel and the distance from the source. The simplest model for a pool fire, one of the most common types of fire in the process industries, assumes a point source and includes a correction factor to allow for the differences in the emissivities of different types of flames:

$$\phi = \frac{F M H_c}{4\pi D^2}$$

where:

ϕ = heat flux, kW m^{-2};
F = fuel factor (see Table 8.10);
M = burning rate, kg s^{-1};
H_c = calorific value of the fuel, kJ kg^{-1};
D = distance from fire, m.

This formula underestimates the flux close to the source (within 2 pool diameters); this can be a problem in designing protective systems and fire-fighting systems.

The other parameters of interest are the surface emissive power (E_{sp}) and the flame temperature. Typical values of F, E_{sp} and flame temperature are given in Table 8.10.

TABLE 8.10
Flux data for fires

Fuel	F factor	E_{sp}, kW m^{-2}	Flame temperature, °C
LNG (pool)	0.2	200	1300
LNG (torch)	0.15	200	1600
LPG (pool)	0.15	150	1300
LPG (torch)	0.3	350	1550
Fuel oil (pool)	0.075	50	1000

The value of E_{sp} for a BLEVE is in the order of 250 kW m^{-2}.

The estimation of the burning rate is difficult and depends upon the latent heat of vaporization and the physical conditions. An approximate formula for a burning pool is given by:

$$M = 0.0046 \times \frac{H_c}{L} \times A_p \times \frac{\rho_1}{3600} \text{ kg s}^{-1}$$

where:

L = latent heat of vaporization, kJ kg^{-1};

A_p = area of the pool, m^2;

ρ_1 = density of the liquid, kg m^{-3}.

This will give an overestimate for very smoky fires.

An alternative model for heat flux calculations assumes that the flame can be represented by a simple geometric shape and that the radiation is emitted from its surface. In this case:

$$\phi = F_g E_{sp} \tau$$

where:

F_g = geometric configuration factor;

E_{sp} = surface emissive power;

τ = atmospheric transmission factor

The geometric configuration factor can be taken as unity if the field of view of the receiving surface is completely filled with the flame, otherwise it is taken as:

$$F_g = \frac{\text{solid angle subtended by emitting surface (in steradians)}}{2\pi}$$

Values of E_{sp} can be found in the literature; Table 8.10 gives some typical values and Table 8.11 gives further data for LNG pool fires.

The value of the atmospheric transmission coefficient τ depends upon the characteristics of the flame, the state of the atmosphere and the path length. There have been a number of attempts to produce values of τ, but for most practical purposes it lies between 0.5 and 1. Hence to be on the safe side $\tau = 1$ unless there is evidence to the contrary.

Flame to metal heat fluxes from pool fires can be up to 65 kW m^{-2} and up to 350 kW m^{-2} for torch flames. Most process equipment cannot withstand fluxes in excess of 12.5 kW m^{-2} for any significant period of time.

TABLE 8.11
Surface emissive powers (kW m^{-2}) for LNG pool fires

Pool diameter, m	Wide angle E_{sp}	Narrow angle E_{sp}
6.1	143	69–150
9–15*	210	185–224
20	153	219
30*	203	—

* LNG pool fires on water

TABLE 8.12
Effect of heat on people

Effect	Thermal dose, (kW m^{-2})$^{1.33}$ s
Start of pain	150
Severe discomfort (possible first degree burns)	250
Second degree burns	1400
Third degree burns	3000

People can withstand between 1 and 2 kW m^{-2} indefinitely (solar radiation can be up to 1 kW m^{-2}). Above these figures there is a time/flux relationship, and the thermal dose is given by:

$$\text{flux}^{1.33} \times \text{time (in seconds)}$$

Figure 8.9 (page 160) shows typical tolerance times for various flux levels and Table 8.12 gives further information on the effect of fire on people.

It must also be remembered that most deaths in fires are due to smoke or fume inhalation rather than burns. No human can tolerate dry air at over 140°C or wet air at 100°C for more than a few seconds and many fires give rise to toxic fumes and gases such as carbon monoxide, phosgene and hydrogen cyanide.

This information is greatly simplified but it does give an indication of the method of assessing the consequences of incidents resulting in fires.

For further information see Reference 9.

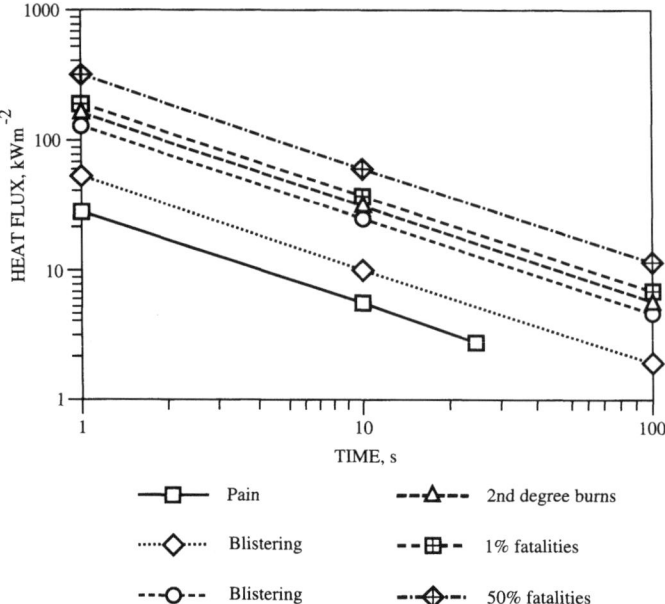

Figure 8.9   Effect of heat flux.

8.5   CONCLUSIONS

Consequence modelling is very complex and must be specific to particular applications. Much work is still needed to refine the models used, although the spread of computational fluid dynamics has done much to improve the state of knowledge as far as dispersion is concerned. Even so, test-work is often needed in order to apply the models with any degree of confidence.

Medical epidemiology is an even more complex and disputed area, though a great deal of work is in progress to assess the long-term effects of dangerous substances. Information in this field can be confusing due to out-of-court settlements where insurance companies often find it cheaper to pay compensation than to prove that their client was not to blame. The main problem is that the most common long-term effect of exposure to dangerous substances (and to ionizing radiation) is cancer and 30% of the population of Western countries die of cancer anyway. This makes acceptable proof or otherwise of the links between particular substances and cancer almost impossible.

However, despite these difficulties, consequence modelling has much to offer safety analysis and decision-makers. Like all aspects of QRA, it is essential that the limitations of the methods, models and data used are fully understood by all concerned.

REFERENCES IN CHAPTER 8

1. Cox, A.W., Lees, F.P. and Ang, M.L., 1990, *Classification of Hazardous Locations* (IChemE, Rugby, UK).

2. Mecklenburg, J.C. (ed), 1985, *Process Plant Layout* (IChemE, Rugby, UK).

3. Pasquill, F., 1961, *Met Mag*, Volume 90 (HMSO, UK).

4. Turner, D.B., 1970, *Workbook of Atmospheric Dispersion Estimates* (USEPA).

5. Briggs, G.A., 1973, *Diffusion Estimation for Small Emissions, ATDL Contribution File No 79.*

6. HSE, 1984, *Guidance Note EH40, Occupational Exposure Limits* (HMSO, UK).

7. Scilly, N., Private communication.

8. Major Hazards Assessment Panel, 1994, *Explosions in the Process Industries*, 2nd edition (IChemE, Rugby, UK).

9. Hymes, I., 1985, Update on the Spanish campsite disaster, *IChemE Loss Prevention Bulletin*, Issue 61.

9.   HUMAN FACTORS

9.1   THE ROLE OF THE OPERATOR

Advances in the application of automation to process plants have resulted in improvements in product quality, productivity and efficiency. Manual control is only used extensively when a particular control parameter is non-sensitive — for example, cooling water flow — or where an infrequent complex sequence has to be carried out — for example, start-up or shutdown. With a highly automated process plant, it is only necessary to inform the operator of those deviations in control parameters which indicate that a dangerous state has occurred or which, if allowed to continue, would lead to a rapidly deteriorating situation. For other parameters information can be provided by the use of status displays. Data loggers can be used for the gathering and recording of general process information.

The operator/machine interface is shown in Figure 9.1.

NATURE OF ERROR

Errors are made in all areas of human activity regardless of training, experience or skill. One of the most important motives for designing fully-automated process plant is the belief that the operator is the weakest link in the man-machine system. Operator reliability depends upon the type of task to be performed:

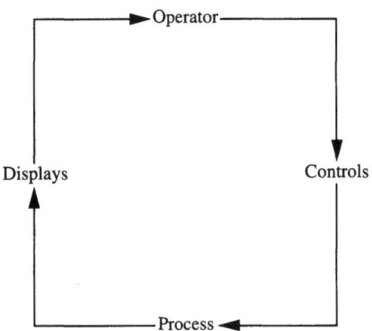

Figure 9.1   Operator/machine interface.

computers are suited for routine control and monitoring whilst operators are more suited to emergency situations which require complex decision-making and action.

Expert systems are being developed which can handle a wide variety of emergency situations without operator intervention, although it will be some time before they can be totally relied upon.

OPERATOR RESPONSE

In hazard analysis a distinction must be made between routine operator action and operator intervention in an emergency. For routine operator action the operator can usually take time and is under no great stress. Safety assessments involve the prediction of the likelihood of errors when the operator is taking corrective action against alarms. The time for corrective action may be short, the operator is liable to be under some stress and so the probability of error is greater.

To assist in the assessment of operator error under such conditions, it is helpful to break the response down into smaller steps:
(1) Operator hears the alarm.
(2) Operator recognizes what the alarm means.
(3) Operator decides what action is required.
(4) Operator performs the action.
Errors are possible at any one or more of these stages.

TYPES OF ERRORS

In order to examine the likelihood of human errors it is first necessary to distinguish between different types of error. They range from a simple error in, say, reading an indicator or operating a control to intentionally carrying out actions which are wrong and which may have disastrous consequences. The former category of error is usually called a slip or lapse and the latter a mistake. There is also a violation which is a deliberate and knowing wrong action; in the ultimate this would be termed sabotage. The distinction between the first two categories is not clear-cut. Operator actions are classified by psychologists into skill-based (those which require repetitive action such as hitting the right button), rule-based (in which the action is dictated by some clearly specified directive) and knowledge-based (which require some thought in order to decide on what action to take). Thus deviations can result from simple physical errors, often made worse by bad design, errors in following rules or operating instructions and errors due to operators misinterpreting situations and formulating incorrect plans of action. Reference 1 provides a comprehensive treatment of the types of human error.

Errors may be due to a wide variety of reasons including:

- operators may not know what to do and pursue the wrong objective (lack of training);
- they may think they know what to do but are wrong;
- they may not want to take a particular action and therefore do not achieve the required objective (self preservation or perhaps lack of motivation);
- they receive contradictory signals and select the wrong information;
- they have received the same warning many times before and, in the past, have been able to ignore it;
- they do not intervene because the tasks may be beyond their physical and/or mental capability.

These causes of error can be eliminated to a large extent by the correct selection and training of operators and by the provision of adequate aids for process control. The operators must also be well-motivated and physically and mentally capable of doing the job.

Even with well-trained and well-motivated operators, however, the occasional error will occur and the probability of such an error depends upon:

- the characteristics of the person;
- the characteristics of the job.

To be able to estimate the probability of operator error, the design of the man-machine interface has to be analysed. The information below can be used to examine both new and existing installations in order to ascertain the risks arising from human error. The machine must be made to fit the operator, not vice versa, and the process interface designed to maximize the strengths of human operators whilst allowing for their weaknesses. This requires some knowledge of the characteristics of human operators and the ways in which they respond.

Operators' strengths
- They have the ability to acquire information from a variety of sources — for example, visual displays, sound and so on.
- They can be versatile — that is, they can make judgements and can consider and combine information to detect problems.
- They can think and react to new situations.
- They can improvise and perform flexible procedures.

Operators' weaknesses
- They can be inconsistent.
- They can be slow to respond to new or complex situations.
- They get tired.

- They have an unreliable 'precise memory'.
- They can develop 'mind set' or 'automatic' response without a full analysis of what is required.
- They are subject to stress.

OPERATOR STRESS

The probability of error is related to the stress under which the operator is working at the time. Under a condition of low stress simulation tests suggest that an operator performs tasks with a basic error rate of 1 in 10^2 to 10^4 operations. Where high stress is involved and the problem is more difficult to define, estimates indicate a 1 in 10 chance of an operator making an error, and the chance is even greater if the time-scale is short. It is therefore important when assessing the probability of human error that an allowance is made for the stress placed on the operator.

The key factors in assessing stress are:
- anxiety;
- time;
- information overload;
- distraction;
- vigilance failures.

Anxiety can be due to perceived or real physical danger resulting from a developing incident on a plant, or it may simply be due to the fear of a reprimand because of erroneous action or poor judgement. Either way it can have a considerable effect upon ability to analyse a situation and come up with the correct solution. The former problem can be partially solved by ensuring that the operator is well protected in an emergency and also by the judicious placing of emergency stops and so on, in order that corrective action can be taken without having to enter a danger zone. The question of fear of reprimand is really a matter of management style. It is important that accident enquiries are made to determine causes rather than to apportion blame. The analyst should therefore look for the existence of a 'no blame culture' for accident and 'near miss' reports.

Higher plant throughputs produce low residence times and rapid responses. The time available for action, therefore, is often less than the reaction time of the operator; the knowledge that the action has to be quick can slow down the decision-making process. Modern nuclear plant is designed so that no operator intervention is needed for up to 30 minutes after a dangerous incident, thus giving time to analyse the situation and decide upon the correct action. The amount of information and the rate at which it is presented often adds to operator stress.

165

Distraction or divided attention occurs when the human brain is required to accept more than one sensory signal at a time. It is very difficult for an operator to attend to two or more unrelated pieces of information simultaneously. Distraction can also be caused by the presence of unauthorized persons in a control room. Error may result from the operator being pre-occupied with outside matters, particularly personal problems, and people should not be left in control of complex plant if they are known to have severe problems of this nature. Control rooms which have degenerated into unofficial mess rooms can be particularly dangerous as operators are very likely to become distracted by the presence of other people.

Individual vigilance is known to fall off rapidly with time, particularly when the job is monotonous. The wandering attention problem is borne out by evidence during the 1939–45 World War, when it was discovered that radar screen operators suffered from a serious deterioration of both attention and reflexes if not relieved every half hour. The safety analyst should always be aware of the increased risk of errors caused by undermanning and double shift working.

It may be good practice with very highly automated systems to include certain routine manual tasks to ensure alertness. Tasks involving varied or complex activities — for example, city driving — seem to be sufficiently stimulating to prevent vigilance failures in most instances although the occasional routine task — for example, monitoring the fuel gauge — may get neglected. On the other hand, accidents due to lack of vigilance on open roads are much greater. The same can be said of many process plants. The boundary between monotonous and stimulating tasks is ill-defined. In general, under a high stress condition, operators cannot be relied upon to act correctly but their response after long periods of inactivity can also be erroneous.

9.2   CONTROL ROOM DESIGN

The probability of operator error depends not only upon the skill, training and stress levels of the operator but also upon the ergonomics of the system. Control panel design has changed dramatically in the last few years due to the use of computers but the same basic principles apply to all types of control rooms.

ENVIRONMENT

To ensure that the health and efficiency of the operator is not affected by the physical working environment, certain minimum standards need to be maintained. The usual factors considered are heat, light, noise, ventilation and vibration. Thus background noises disguise a sound signal if too obtrusive and bad lighting and reflections can make instruments and visual display units difficult to read. Operators are more likely to maintain peak performance when working conditions are good.

DISPLAYS

The layout of displays on the control panel should represent the functional arrangement of the process plant in order to facilitate the building of a mental model of the plant by the operator. Psychologists make great play of the importance of operators having a correct mental model of the plant that they are controlling. It is easier to build a mental model with the old style mimic diagram control panels but much can be done by the use of overview VDU displays and static diagrams.

ALARMS

For operator understanding, status displays and alarms should be provided so that the information can be presented in a manner which is easy to interpret. Multiple and duplicated alarms can cause confusion and present the operator with too much information to digest at once. Furthermore, spurious alarms such as those resulting from maintenance work can confuse the operator in times of high stress.

The alarm system should direct the operator to the disturbance that caused the problem. The operator should not be left to deduce this information purely from the pattern of alarms and timing. Thus the requirement is for information to give a presentation which is easy to use and which helps the operator to diagnose the root cause of a process deviation, so that the appropriate action can be taken.

Alarms should be grouped with the controls and instrumentation for the section of plant to which they apply. Subsidiary alarms should be suppressed in an emergency situation. Up to three categories of alarm, distinguished by sound and colour, can be used to distinguish impending disaster from minor malfunctions.

INSTRUMENTATION AND CONTROLS

The information presented must be reliable — that is, indications must come from actual devices, not from the signals to them. Thus valve positions must come from limit switches and rotational indicators should be provided on driven equipment; it is not sufficient to rely on motor running lights. One of the main reasons why the true cause of the core melt-down at Three Mile Island nuclear power plant in the USA was not diagnosed sooner was that the status indication of the key valves came from the operating switches and not as feedback from the valves themselves. The operators thought that a particular valve was closed when in fact it had stuck open.

Those instruments and controls which are of special importance — for example, critical or frequently used instruments and controls — should be placed in a prominent and accessible position and should be clearly distinguishable from minor controls. It is essential that controls respond in the way operators expect — for instance, if a control is turned clockwise or moved upwards the controlled variable should increase. This can vary from culture to culture even within the Western world, the domestic light switch being a good example. Gauges should if possible have linear scales and the graduations be clear and readily understood. Great care must be taken to avoid confusion with abbreviations.

An excellent guide to the review of control room design and layout, and hence an indication of the likelihood of error resulting from poor design, is given in Reference 2.

GENERAL GUIDELINES

In order to assess the probability of human error the safety analyst should consider the following points:
- sound ergonomics is important. Ensure that instruments and alarms are easily read and controls are within easy reach;
- controls should respond in the expected manner;
- important controls should, if possible, be fail-safe;
- similar plant should, as far as possible, have similar controls;
- design alarms to enable operators to diagnose fault conditions. Avoid multiple alarms if possible;
- do not use alarms for information in normal operation;
- plant status indications must be by active feedback from the plant items concerned;
- avoid work situations which are monotonous;
- do not give operators conflicting information;
- give operators time to think;
- give operators good training, test them frequently and re-train them if necessary.

9.3   HUMAN ERROR ASSESSMENT METHODS

Over the years there have been many methods suggested for the assessment of human reliability; one of the best summaries of the techniques is given in Reference 3. This work gives details of eight quantitative or semi-quantitative techniques. In addition the traditional Hazop techniques can, with the use of suitable guide words, also be applied to human reliability. There is much debate as to the applicability of the techniques to particular problems and several comparative studies[3,4] have been made.

ERROR PREDICTION

The various classes of error were discussed in Section 9.1 (page 163). Only skill-based errors can be quantified with any degree of certainty; thus data is available — for example, on the likelihood of an operator hitting the wrong button — and reasonable correlations have been made for many error situations. Rule-based errors are harder to quantify since they involve more complex mental processes and are thus more likely to be affected by the mental and physical state of the individual concerned. Because knowledge-based decisions are even more complex and involve the operator's mental model of the plant or equipment concerned, quantification is very difficult — some analysts would say impossible.

PREDICTION TECHNIQUES

The literature contains a large number of techniques for human error prediction; amongst the more common are:
- Technique for Human Error Prediction (THERP);
- Success Likelihood Method (SLIM);
- Paired Comparisons (PC);
- Absolute Probability Judgement (APJ);
- Human Error Assessment and Reduction Technique (HEART);
- Technica Empirica Stima Operatori (TESCO);
- Human Cognitive Reliability (HCR);
- Influence Diagram Approach (IDA).

Most techniques follow the same generic pattern. One of the most used techniques is the Technique for Human Error Prediction (THERP). The assessor first analyses the task into a series of events and each event into a set of two possible outcomes. Thus a branching chain of actions is produced in which some paths are successful and others failures. The assessor then determines the probability of error for each branch of the tree. This requires data. One good source is the THERP Handbook[5] which gives numbers for a wide range of operations, qualified by upper and lower uncertainty bounds.

This data is for 'normal' conditions and the basic error rate must then be corrected for a number of factors. The first correction is for 'dependency' — that is, the fact that if a person makes one error, that person may be more or less likely to make another or so alter the situation so that a different person is more or less likely to make another error. This is very much a task for a skilled analyst.

169

The basic error rate is then modified for any abnormal features such as operator stress, poor operating conditions, inadequate training or supervision, poor motivation and so on. With regard to operator stress, consideration must be given to factors such as the time available to make decisions and perform actions, the emotional state of the operator and the risk perceived by the operator. This is usually done using Performance Shaping Factors (PSFs).

The next step is to take the whole chain of probabilities for the different steps in the task and calculate the overall probability of failure. This is termed the Human Error Probability (HEP) which can then be combined with other causes of failure to give a complete quantified failure rate. Because the method is highly subjective in some areas and because there is a paucity of valid data, it is desirable at this stage to carry out a sensitivity analysis and, if necessary, examine some of the data and assumptions used in more detail.

The great advantage of THERP is its consistency with methods used for equipment reliability assessment, but it is time consuming and requires a skilled analyst. Success Likelihood Method (SLIM) uses a combination of available data and expert judgement, the judgement being 'calibrated' by reference to known data. PSFs are an important part of SLIM and the method of combining them is different from that used in THERP. The Paired Comparisons (PCs) technique may be used with SLIM or on its own. It involves asking teams of experts to make simple judgements between pairs of likely outcomes to decide which is the more probable. The deliberations of the experts are then combined to give a scale of error likelihood. Absolute Probability Judgement (APJ) uses either a single expert or teams of experts to come up with probability figures using their own and shared expertise.

Human Error Assessment and Reduction Technique (HEART) is a newer technique which is based on a review of human factors literature and experimental data on human performance. The technique defines a set of generic probabilities for different types of tasks and uses them as the starting point for quantification. The analyst looks for HEART error producing conditions (EPCs) and multiplies the basic HEP by the appropriate EPC factor, thus increasing the HEP. The technique also suggests a number of error reduction strategies that can be applied to reduce the likelihood or the impact of the error. HEART is also essentially a subjective approach which requires skill through familiarity to yield any measure of confidence.

Another version of this technique is Technica Empirica Stima Operatori (TESCO) which uses three generic task types adjusted by four PSFs based on time available, extent of training/experience, anxiety and ergonomic quality.

The influence of timing on operator reliability is covered by two techniques — Human Cognitive Reliability (HCR) and Time Reliability Correlation

(TRC). Both have been developed essentially for the nuclear industry and they use a logic tree approach and a set of modifiers based on operator experience, stress and interface quality. The modifiers (TRCs) are now based on data obtained from nuclear plant simulators and are categorized according to the three behaviour types (skill, rule and knowledge). The TRCs take into account such factors as hesitancy due to workload, uncertainty and decisional conflict.

Decision analysis theory has been applied to human reliability in the form of the Influence Diagram Approach (IDA). This approach is based on the principle that human reliability is determined by the combined influence of such factors as quality of information available, organizational and individual factors. Such factors are, in turn, modified by lower level influences. The information is presented in the form of an Influence Diagram and numerical estimates used to weight direct estimates of error probability.

In addition to all these techniques a large human error database (HED) produced by the US NRC (WASH 1400)[6] is available. This can be used in a manner similar to THERP but without the various modifiers. A number of other techniques appear in the literature but they are all similar to one or other of those described here.

COMPARISON OF METHODS

The analyst is faced with a very difficult choice in selecting the technique most suitable for a particular purpose. A number of comparisons have been made and guides produced. Tables 9.1 and 9.2 on page 172, taken with permission from Reference 3, give a good summary of one set of evaluations. The longest established technique, THERP, is still the most commonly used and is generally accepted by regulatory authorities. Its main weakness is that it does not really handle knowledge-based tasks, and it is expensive to use. HEART copes with knowledge-based tasks and is much cheaper to apply, though it has yet to gain the same degree of acceptability as THERP. Both techniques can be applied by trained analysts without having an expert knowledge of the system concerned. Most of the other techniques require experts skilled in the relevant approach and are therefore more difficult to apply.

The only technique that has been subjected to a very thorough validation process is THERP. The results[4] were rather disappointing and attempts to analyse the same problem by different techniques show a wide spread of results. This work underlines the need to pick the most suitable technique for the problem and the importance of applying sensitivity analysis to critical areas.

TABLE 9.1

Evaluation of Human Reliability Assessment techniques[†]

	APJ	PC	TESCO	THERP	HEART	IDA	SLIM	HCR
Accuracy	M	M	L	M	M	L	M	L
Validity	M/H	M	L	M	M	M	M	L
Usefulness	M/H	L/M	M/H	M	H	M/H	H	L/M
Use of resources*	M	L/M	H	L/M	H	L/M	L/M	M
Acceptability	M	M/H	L	H	M	M	M/H	L/M
Maturity	H	M	L	H	L/M	L/M	M/H	L

H = high; M = moderate; L = low

* A ranking of high in this criterion indicates effective use of resources

[†] Reproduced from *Human Factors and Decision-Making*, Sayers, B.A. (ed), 1988 (Elsevier Science).

TABLE 9.2

Human Reliability Assessment techniques selection matrix[†]

	APJ	PC	TESCO	THERP	HEART	IDA	SLIM	HCR
Use for simple tasks	Y	Y	Y	Y	Y	Y	Y	N
Use for knowledge-based, abnormal tasks	Y	Y	Y	N	Y	Y	Y	Y
Use for misdiagnosis which makes things worse	Y	Y	N	N	N	Y	Y	N
Qualitative recommendations possible?	Y	N	Y	Y	Y	Y	Y	Y
Sensitivity analysis possible?	N	N	Y	Y	Y	Y	Y	Y
Calibration data required?	N	Y	N	N	N	N	Y	N
Experts required?	Y	Y	N	N	N	Y	Y	N

9.4 APPLICATION OF HAZOP TO HUMAN RELIABILITY

The traditional Hazop study technique can, with some modifications, be applied to human reliability situations. The usual process parameters are replaced by two new words:

- INFORMATION;
- ACTION.

and the guide words slightly modified as follows:

- MORE;
- LESS;
- NO;
- WRONG.

The parameter INFORMATION applies to information available from displays, procedures, previous training, experience, communications and any other source which the operator may use. The parameter ACTION refers to operator response. Errors in ACTION may be in terms of incorrect selection or incorrect execution of a response.

The method can be illustrated by a very simple example taken, with permission, from Reference 3.

Deviation
NO ACTION

Causes
Control cannot be accessed
Necessity for action not perceived
No information to act on
No operator present
Operator distracted
Communication failure
Action too late
Assume another person has acted
Insufficient time to complete
Failure to restore automatic control
No supervision/checking/testing

9.5 DATA ON OPERATOR RELIABILITY

There is a wide spread in the published information on operator reliability and any figures quoted must be carefully examined for applicability to a particular situation. Table 9.3 provides some typical figures obtained from a wide range of situations which at least give a guide to what may be expected of operators.

Table 9.4 gives an indication of the problems associated with acting inside a very short time span; the figures in some ways simplify the problem because if an operator has to wait before carrying out a certain action, there is always a danger that it will be forgotten about or that the delay will give rise to anxiety which reduces performance.

The probability of error is greatly increased (by as much as an order of magnitude) if the operator perceives a great personal danger — for example, a risk of explosion. Figure 9.2, taken from Reference 5, offers further data on the probability of operator failure.

TABLE 9.3
Data on operator error

Type of operation	Error rate
Complicated; non-routine	1 in 4
Non-routine; other duties at the same time	1 in 10
Routine; requires care*	1 in 100
Routine; simple	1 in 1000
Simplest possible action	1 in 10,000 to 1 in 100,000

* An example of a 'routine; requires care' situation is keying in a telephone number.

TABLE 9.4
Operator reaction probabilities

Time for operator action	Operator aids	Probability of error
1 s	Alarm	1
10 s	Alarm	10^{-1}
60 s	Alarm	10^{-2}
5 min	Alarm	10^{-3}
10 min	Alarm	10^{-4}

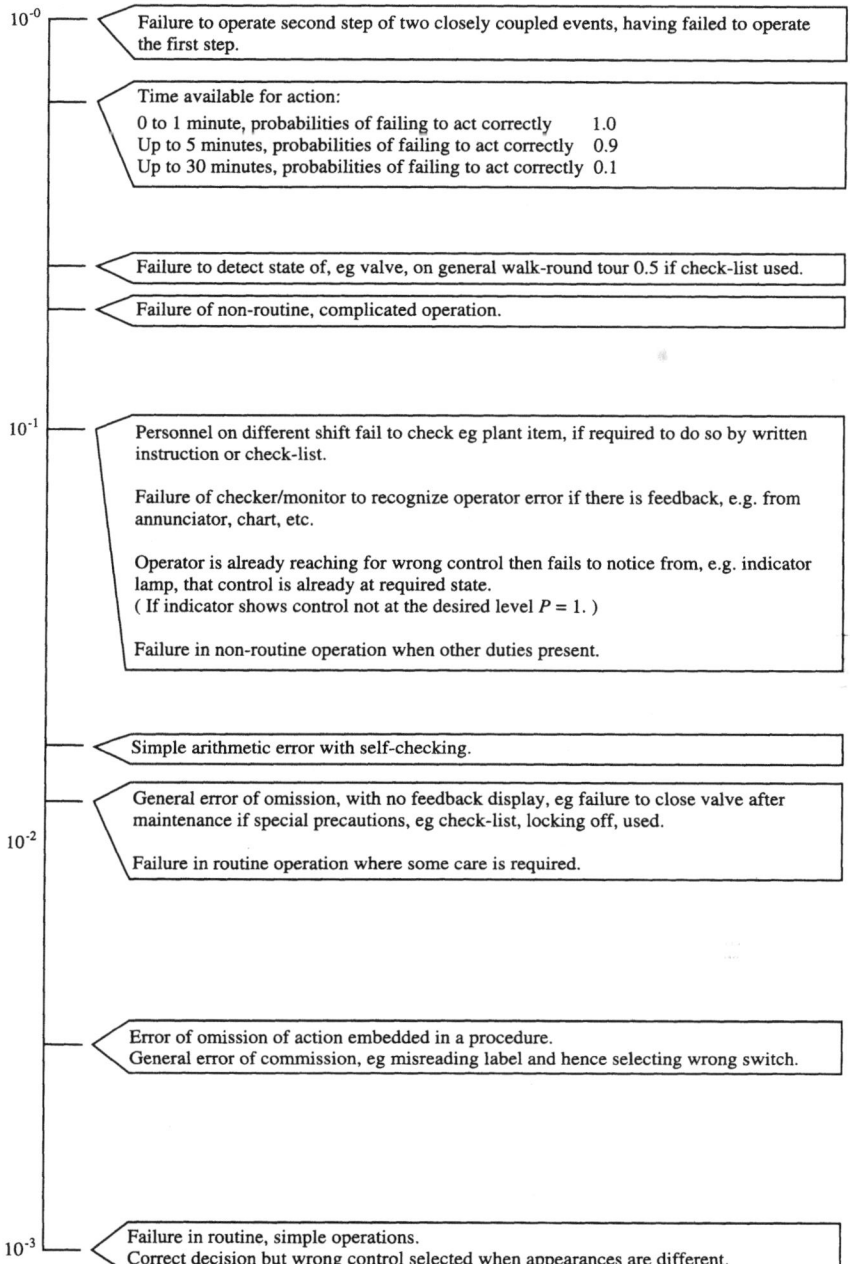

Figure 9.2   Operator reliability data.
Reproduced from Swain, A.D. and Gutterman, H.E., *Handbook of Human Reliability Analysis, NUREG–CR/1278* (US Nuclear Regulatory Commission, Washington, DC).

9.6   THERP — AN EXAMPLE

The situation is a steadily running plant. The analyst requires an estimate of the probability that the control room operator will not respond within, say, one minute to an alarm when attention is distracted by a second alarm. When the operator hears an alarm it is acknowledged by means of a two-stage 'accept' system. The operator first cancels the audible alarm and after interrogating the panel presses a second accept button which changes a flashing visual signal on the panel to a steady signal. The steady signal will remain until the fault is rectified.

Experience for this particular plant indicates that for a single alarm the probability of the operator not recognizing an alarm after cancelling the audible signal is 10^{-4}, and the probability of the operator not taking action straight after recognizing the alarm is 10^{-5}. Experience also indicates that if the fault is not addressed within one minute, the probability of it being forgotten rises from 10^{-4} to 10^{-3}. Thus if a second alarm sounds, the probability of the operator failing to recognize the second alarm after cancelling the audible signal is 10^{-3} because the operator will still be thinking about the first alarm. The probability of the operator rediscovering the alarm after, say, one hour, becomes 0.05 — that is, the probability of the operator noticing the visual warning in a routine scan of the panel. This figure can be even lower if there are always a number of visual warnings on the panel due to long-term faults or maintenance operations. Hence the importance of ensuring that the plant is not run continuously with visual alarms showing on the panel.

The sequence is shown in Figure 9.3, taken from Reference 5.

If the probability of a second alarm occurring within one minute of the first alarm is 10% (rather high but used for illustration purposes only), the probability of failure of the operator to respond correctly to both alarms can be calculated as follows:

$$
\begin{aligned}
F_1 &= 0.95 \times 10^{-4} & &= 0.095 \times 10^{-3} \\
F_2 &= 0.9999 \times 0.9 \times 10^{-5} & &= 0.008999 \times 10^{-3} \\
F_3 &= 0.9999 \times 0.1 \times 0.001 \times 0.95 & &= 0.09499 \times 10^{-3} \\
F_4 &= 0.9999 \times 0.1 \times 0.999 \times 10^{-5} & &= 0.000999 \times 10^{-3} \\
F_5 &= 0.9999 \times 0.1 \times 0.999 \times 0.99999 \times 10^{-3} \times 0.95 & &= 0.094889 \times 10^{-3} \\
F_6 &= 0.9999 \times 0.1 \times 0.999 \times 0.99999 \times 0.999 \times 10^{-5} & &= 0.000999 \times 10^{-3} \\
\text{TOTAL} & & &= 0.2959 \times 10^{-3}
\end{aligned}
$$

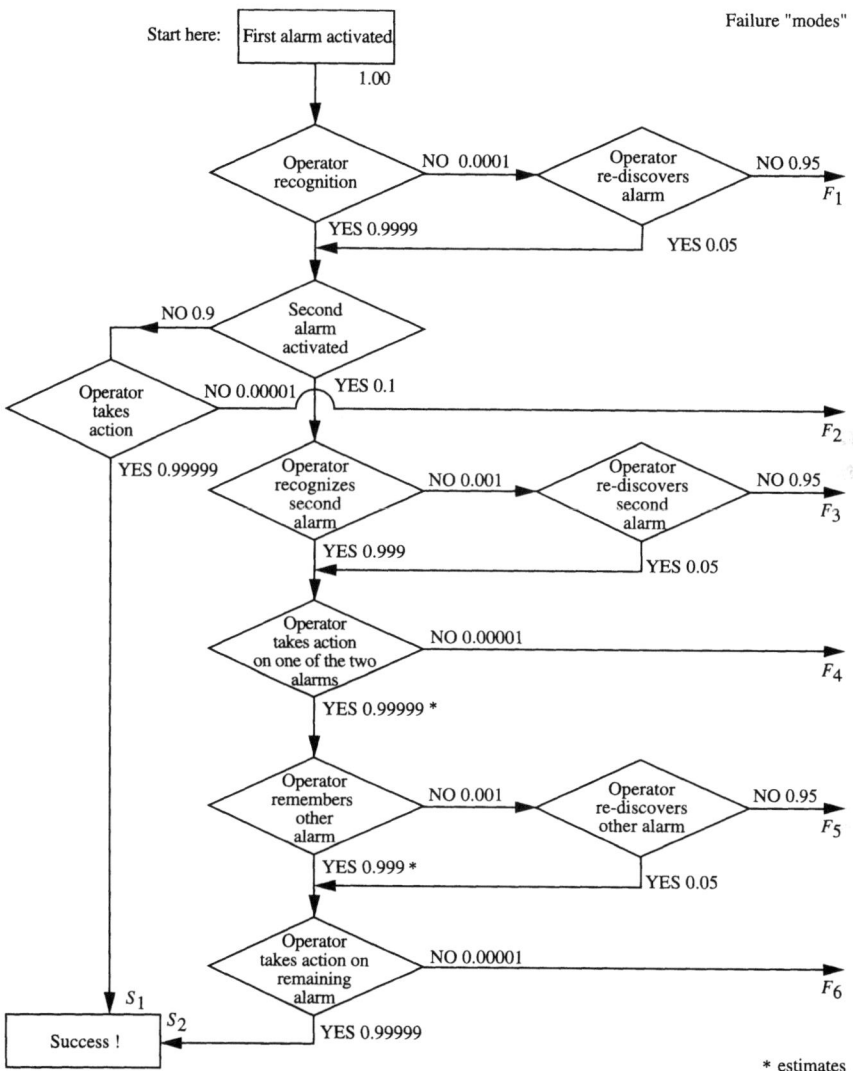

Figure 9.3   THERP flow diagram.
Reproduced from Swain, A.D. and Gutterman, H.E., *Handbook of Human Reliability Analysis, NUREG–CR/1278* (US Nuclear Regulatory Commission, Washington, DC).

177

9.7   CONCLUSIONS

Total elimination of human error will never be possible and it will be many years, if ever, before expert systems can totally replace the human operator. In the meantime use must be made of the science of ergonomics to ensure that everything possible is done to enhance the strengths of human operators whilst at the same time allowing for the weaknesses. There are many methods now available to quantify human error and to provide data for QRA but great care is needed to ensure the selection of the most appropriate technique. Human error data is even more prone to misuse and misinterpretation than equipment error data. None the less the understanding of human error has now advanced far enough for predictions of operator reliability to be made with some degree of confidence.

REFERENCES IN CHAPTER 9

1.   Rasmussen, J., 1979, *On the Structure of Knowledge. A Morphology of Mental Models in a Man-Machine Context* (RISO–M–2192, Roskilde, Denmark).

2.   *Guidelines for Control Room Design Reviews, NUREG–0700* (US Nuclear Regulatory Commission, Washington).

3.   Sayers, B.A. (ed), 1988, *Human Factors and Decision-Making* (Elsevier Applied Science, London, UK).

4.   ACSNI Study Group on Human Factors, 1991, *Second Report* (HMSO, London, UK).

5.   Swain, A.D. and Gutterman, H.E., *Handbook of Human Reliability Analysis, NUREG–CR/1278* (US Nuclear Regulatory Commission, Washington, DC).

6.   *WASH 1400 Reactor Safety Study, NUREG–75/014* (US Nuclear Regulatory Commission, Washington).

10. CONCLUSIONS

All engineers have a duty to ensure that the plants which they design and operate are as safe as is reasonably practicable. Compliance with this duty requires a proactive approach at all times. A sound safety culture must first be established throughout the organization and use made of the most appropriate tools of safety assurance at all stages in design and operation.

The most important rule must always be 'inherent safety is better than engineered safety', 'what you haven't got can't leak'. The most elaborate safety devices, validated by the best available techniques of safety analysis, will never reduce to zero the risk due to the escape of a noxious substance, whereas replacing a noxious substance by a more benign one could well eliminate that risk altogether.

It has never been possible to remove risk from human activity and it never will be. All that engineers can do is to control the risks associated with their activities and ensure that the risks are commensurate with the benefits gained. This book has attempted to put the risks associated with process plant into perspective and to describe the various methods available to answer the following questions:

- what can happen?
- how often is it likely to happen?
- what are the consequences?
- are the resultant risks acceptable?

Whilst the answer to the last question may ultimately be the responsibility of the politician rather than the engineer, it is essential that all engineers understand not just the mathematics of risk analysis but also the philosophical and moral arguments involved.

11.   PROBLEMS

INTRODUCTION

Here are some problems to test your grasp of what you have read. The answers are found on pages 194–206. Problems marked with an asterisk are based on questions taken from the University of Cambridge Chemical Engineering Tripos; the consent of the University is duly acknowledged.

No responsibility can be taken for the accuracy either of the data used or of the answers given. The problems are chosen to illustrate certain points and should not be taken as indicative of practices in the industries concerned.

PROBLEM 1 (answer on page 194)

(a)   Using a true dice, what is the probability of:

(i)   throwing a total of exactly seven in two throws?

(ii)   throwing a total of more than three in two throws?

(iii) not throwing a one or a two in two throws?

(iv) throwing a 3, 4, 5 or 6 at least once in two throws?

(b)   If event A happens twice per year and lasts on average for 9 hours, and event B happens 14 times per year and lasts on average for 5 hours, what is the probability of:

(i)   states A and B existing coincidentally?

(ii)   either state A or state B or both existing?

(iii) neither state A or state B existing?

PROBLEM 2 (answer on page 194)

Figure 11.1 shows a trip system on a distillation column vapour line which cuts off the steam to the reboiler if the column pressure rises above a preset value. Using the data given below, calculate the fractional dead time (*fdt*) of the system:

(a) on the basis of a proof test every 4 weeks;

(b) on the basis of a proof test every *T* years.

Hence calculate the optimum proof test interval and the associated *fdt*.

Component	Failure rate, yr^{-1}
Impulse line	0.09
Pressure switch	0.1
Relay	0.01
Solenoid valve (SOV)	0.1
Trip valve	0.1
Air line blocked	0.01

- Disarm time for testing: 1 hour.
- Probability of test causing trip to remain dead until next test: 1 in 1000 tests.

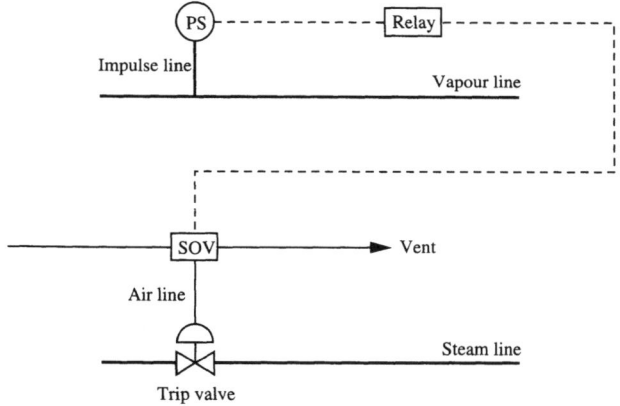

Figure 11.1 Distillation column vapour trip system.

PROBLEM 3 (answer on page 194)

Figure 11.2 shows a cooling system for a hydrocarbon gas. The heat exchanger is bypassed if the inlet temperature (as measured by T1) is below 100°C and the cooling water pump is started if the outlet temperature (as measured by T2) exceeds 100°C. Independent control units take the temperature sensor outputs and send signals to the diverter valve and the pump.

Using the data given below, produce a fault tree for the gas temperature exceeding 100°C and estimate the probability of failure in one operating year of 8000 hrs. Comment on the result and suggest improvements to the system.

Item	Mean failure rate (per 10^6 hours)
Temperature sensor	12
Diverter valve	95
Heat exchanger (all causes)	99
Centrifugal pump (all causes)	120
Control units	25

The probability of there being no sea water due to line blockage, inadvertent valve closures or other reasons can be taken as 0.01 and the probability of the hydrocarbon inlet temperature exceeding 100°C can be taken as 0.5. Note that the failure rates are such that the approximate formulae cannot be used.

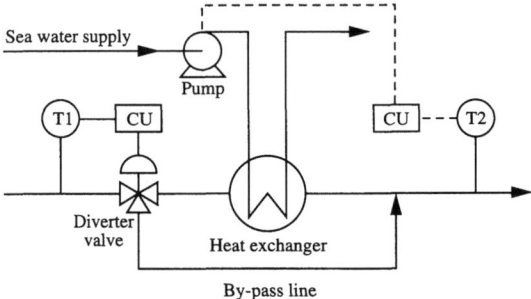

Figure 11.2 Hydrocarbon cooling system.

PROBLEM 4* (answer on page 196)

A reactor effecting an exothermic reaction is at risk of thermal runaway in the event of coolant failure. Its protective trip system is intended to open a dump valve which empties the reactor if low coolant flow or high reaction temperature is detected. Draw a fault tree which summarizes the failure logic analysis given below. Calculate the approximate frequency of the runaway reaction. To which of the primary failures is it most sensitive?

FAILURE LOGIC ANALYSIS

Runaway reaction occurs if cooling water failure occurs whilst the protective system is inoperative. Cooling water failure can occur because of pump failure, line blockage or an exhausted water supply. The protective system may be inoperative when either the shutdown system fails because the dump valve fails shut, or because the detection system fails.

Failure	Failure rate
Pump failure	$0.2 \ yr^{-1}$
Line blocked	$0.01 \ yr^{-1}$
Supply tank empty	$0.1 \ yr^{-1}$
Dump valve fails shut	0.001/demand
Low flow trip failure	0.01/demand
High temperature trip failure	0.01/demand

PROBLEM 5* (answer on page 196)

The highly active waste liquor (HAL) arising from the reprocessing of nuclear fuel continues to generate heat due to the decay of fission products. Prior to vitrification it is stored in stainless steel tanks equipped with multiple cooling coils and a jacket. If the cooling system completely failed, the liquor would eventually boil, leading to the possible release of radioactive material.

The cooling water is normally recirculated through mechanical draught cooling towers and make-up water comes from a nearby lake.

In the event of a complete failure of the cooling tower based supply, a connection can be made to enable cooling water to be pumped from a local river using portable fire pumps. No other pipework is required for this operation; a direct connection is made to the jacket and coils.

A spare tank in always kept on standby so that, in the event of cooling coil or jacket failure, the contents can be transferred via a valve and a diverter using a steam powered ejector (no electricity needed). This complete standby system is tested once every month.

Using the data given below, produce a fault tree and estimate the annual frequency of boiling of the HAL. If the end result is unacceptable, what steps would be recommended to improve the system?

Item	Failure rate, yr^{-1}
Tank cooling (jacket and coil)	1×10^{-4}
Tank valve (HAL outlet)	4×10^{-4}
Water pipework (cooling system)	1×10^{-4}
Spare tank	1×10^{-4}
Transfer steam supply	2×10^{-7}
Steam pipework	1×10^{-3}
Ejector	2×10^{-3}
Diverter	4×10^{-3}
Cooling tower system (mechanical)	4×10^{-4}
Normal cooling water supply (lake)	1×10^{-5}
Electricity supply	1×10^{-6}

Item	Unavailability on demand
Fire pump	4×10^{-3}
River water	1×10^{-3}

Item	Error probability
Connection error (fire hoses)	1×10^{-3}

PROBLEM 6* (answer on page 197)

Define the meanings of the following terms used in estimates of the reliability of trip systems:

(a) hazard rate H;

(b) demand rate δ;

(c) failure rate λ;

(d) fractional dead time ϕ;

(e) proof test interval τ.

Under what circumstances is the relationship '$H = \lambda \delta t / 2$' valid?

A particular trip has a failure rate of 0.7 faults/yr on a duty where the demand rate is 0.5 demands/yr. When it is being tested, the trip is disarmed for 1 hour, and the probability that the trip is left isolated after testing is 0.001. Calculate the optimum testing interval for the trip in order to minimize the hazard rate; what is the expected hazard rate if this optimal testing interval is adopted?

Since this test interval is unacceptably small, consideration is being given to employing two trips in parallel. Assuming that the fractional dead time for two trips in parallel is approximately $\dfrac{4\phi_1^2}{3}$ where ϕ_1 is the fractional dead time for a single trip due to failure only, estimate the testing interval for the two trip system to give the same hazard rate. It can also be assumed that the fractional dead time is dominated by the probability of failure between tests.

Discuss the circumstances under which $\dfrac{4\phi_1^2}{3}$ would underestimate the actual fractional dead time of a parallel two-trip system and the strategy for timing and trip disarmament that should be adopted during testing.

PROBLEM 7* (answer on page 197)

A process plant has a twin channel protective system with an individual failure rate of 0.4 yr^{-1}. A trip of either channel will return the plant to a safe condition. The time to test (and repair if necessary) each channel is half a day and there is a probability of 1 in 1000 that a testing error puts the channel out of service until the next test. The same test interval is used for each channel. Dependent failures may be neglected and λt (the cumulative probability that a single channel will fail in time t) can be taken as very much less than 1.

Show that the optimum test interval is about 11 days. Comment on the practicality of this figure and calculate the fractional dead time for this test interval.

PROBLEM 8* (answer on page 198)

Figure 11.3 shows the design of a reactor quench system. The purpose is to deliver a prescribed quantity of a hydrocarbon solvent to the reactor when the latter goes out of control, and thereby cool the reactor contents to a safe temperature. The reactor goes out of control on average once every two years.

Construct fault trees for the top events (a) 'tank ruptures due to overfilling' and (b) 'reactor explodes', assuming in both cases that the tank initially contains its proper charge of solvent. Calculate the hazard rate in each case using the data given below.

If each shift has 10 operators with 3 shifts working 8 hours on any one day, calculate the FAR for these men if a reactor explosion results in one of these men being killed. FAR can be taken as the number of workers killed per 10^8 working hours. If the company concerned has a target FAR of 0.4 for any single hazard, suggest possible alterations to the control system in the light of the calculated result.

Failure rate	
Control valve	0.6 yr^{-1} fails stuck in position
Valve positioner	0.44 yr^{-1} fails stuck in position
Level transmitter	1.70 yr^{-1} equal chance of failing high or low
Level controller	0.29 yr^{-1} equal chance of failing high or low
Level switch	0.34 yr^{-1} equal chance of failing high or low

- Each device is proof tested at monthly intervals.
- The overflow has a fractional availability of 0.95.
- The process trip is 100% reliable.

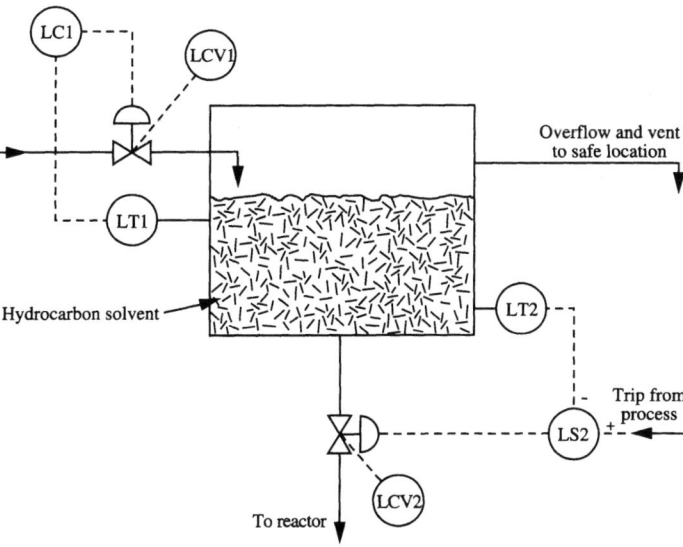

Figure 11.3   Reactor quench system.

PROBLEM 9* (answer on page 199)

Figure 11.4 depicts the separation train on an offshore gas platform. A two-phase mixture of hydrocarbon gases and condensate enters the separator which is designed to separate the liquid from the gas. The gas stream passes on to a knock-out drum and hence to the suction of a compressor. Construct a fault tree with a top event 'liquid enters compressor' and estimate the probability of the top event occurring in any one year.

The system is protected by an emergency shutdown (ESD) valve operated by extra high level in either the separator or the knock-out drum. How does this value change if an additional high level trip is added to the KO drum? Neglect any common mode effects.

What would be the payback time of the additional trip on the basis of (a) damage to the compressor alone and (b) the supposition that the damage would cause a partial shutdown of the platform?

Level control system failure probability	0.2
Level switch (LSHH) and ESD system	0.3 yr^{-1}
Proof test interval (LSHH and ESD)	Monthly
Level switch and ESD system cost	£10,000
Compressor repair time	14 days
Compressor repair cost	£278,000
Loss of production costs	£10m per day

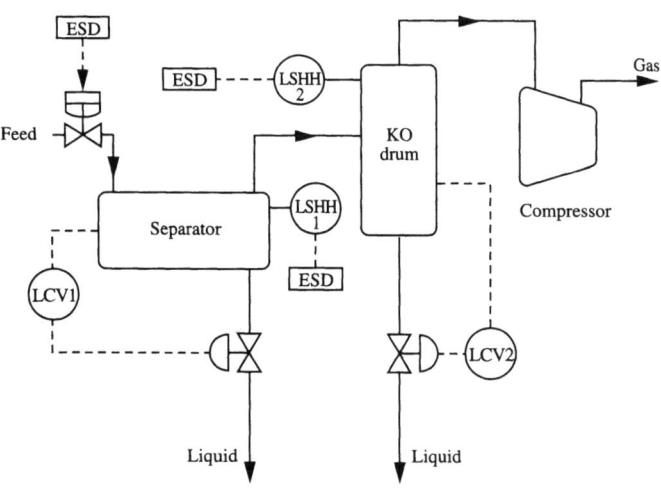

Figure 11.4 Separation on an offshore gas platform.

PROBLEM 10* (answer on page 200)

An LPG storage tank installation is sited close to a railway line on which trains pass carrying fuel oil to a power station. If a train derails it may either plough directly into the LPG storage installation or it may overturn with a consequent possibility of the fuel oil catching fire. The fire may cause the LPG installation to explode.

Use the data given below to estimate the frequency of explosion of the LPG storage installation.

Number of fuel oil trains per year	1000
Derailment frequency per km travelled	0.4×10^{-6}
Length of average fuel oil train	0.5 km
Length of LPG installation next to railway line	0.7 km
Probability that a derailed train overturns	0.5
Probability that an overturned train catches fire	0.1
Probability that fire engulfs LPG tanks causing explosion	0.2
Probability that a derailed train hits the LPG installation causing an explosion	0.05

PROBLEM 11 (answer on page 201)

Discuss the use of 'event trees' as a method of safety analysis in the process industries.

Figure 11.5 shows a distillation unit handling flammable material operating at an elevated pressure. The column bottoms system includes a liquid cooling train and the liquid is discharged to a tank which is not designed to withstand full column pressure. Under normal operation, a liquid seal is maintained in the column base and the bottom product is let down through a control valve linked to liquid level in the column. In the event of failure of the control system and its associated back-up, it is possible for high pressure gas to break through into the low pressure system. Using the probability data given below, construct an event tree for the consequences of breakthrough and estimate the probability of fatal injury.

The plant is run continuously and operated by a single process worker on a shift basis. Each shift worker will work 250 eight hour shifts per year. Calculate the maximum frequency of liquid breakthrough to satisfy operating company safety criteria.

Cooling system fractures	0.02
Vapour cloud ignites	1.00
Operator present in vapour cloud area	0.80
Operator escapes before ignition	0.25
Operator burned but survives	0.50

- The operating company has a target FAR of 0.4 for a single incident.

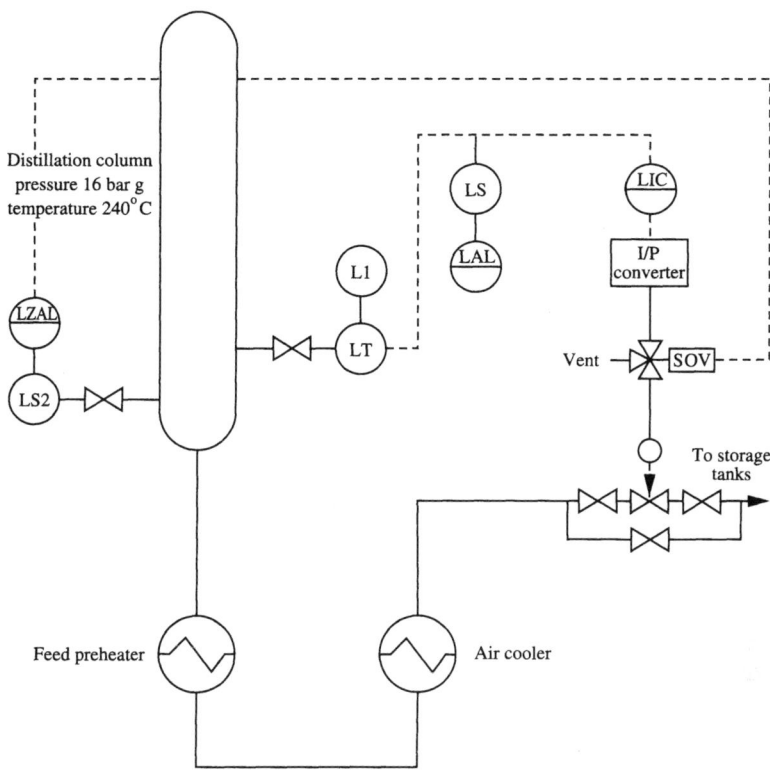

Figure 11.5   Distillation of flammable material at high pressure.

PROBLEM 12 (answer on page 202)

Carry out a failure mode and effect analysis (FMEA) for the system described in Problem 11. Using the failure rate data below and applying 'analyst's judgement' perform a FME criticality analysis to determine the weak link(s) in the system.

Failure frequency, occ yr^{-1}	
Isolating valve shut	0.02
Impulse lines blocked	0.03
Level probe stuck	0.05
Total transmitter failure	0.5
Level switch stuck	0.1
Level control total failure	0.3
Level control loss of output	0.1
Level control set point error	0.02
I/P converter output error	0.1
Solenoid valve stuck	0.3
Control valve fail	1.0
By-pass open	0.1
Level alarm fail	0.1

PROBLEM 13 (answer on page 206)

A safety study on a plant using petroleum gas as a fuel has estimated the maximum possible leak to be 10 m^3 hr^{-1}. If the composition of the gas can be taken as 40% v/v propane and 60% v/v butane, calculate the minimum ventilation rate required to ensure that the concentration of flammable gases does not exceed 25% lower explosion limit (LEL).

If the size of the building is 10 m by 20 m by 15 m high and the gas can be considered to be perfectly mixed in the atmosphere, how long has the operator got to restore power supplies in the event of ventilation system failure?

PROBLEM 14 (answer on page 206)

An explosives plant has a maximum inventory of 10 tonnes of a commercial explosive with a heat of combustion of 6000 kJ kg^{-1} and the same efficiency ratio as TNT. Calculate the explosion overpressure in a site building 50 m from the plant and at a proposed housing development 1 km from the plant.

Give an estimate of casualties and damage at each location. Assume that the average occupancy of the site building is 5 and the housing development would be 500. If the frequency of an explosion is estimated at 10^{-4} yr^{-1}, comment on the figures.

PROBLEM 15 (answer on page 206)

A road tanker containing fuel oil is involved in a traffic accident. The tank is ruptured and a burning pool of fuel is formed on the road 3 m in diameter. How close can the emergency services approach without protective clothing? How long has an unprotected rescuer got to move an injured person lying 6 m from the centre of the fire without risking incurring first degree burns?

Heat of combustion of fuel oil	43.5 MJ kg^{-1}
Latent heat of vaporization	294 kJ kg^{-1}
Density of fuel oil	875 kg m^{-3}

ANSWERS TO PROBLEMS

The answers are often given to more significant figures than can really be justified by either the accuracy of the data provided or by the validity of the empirical equations used. This fact must be borne in mind when quoting answers to any quantitative safety analysis and the appropriate factors of safety and/or uncertainty must be applied.

Problem 1

(a)   (i) 1/6, (ii) 11/12, (iii) 4/9, (iv) 8/9

(b)   (i) 0.000016, (ii) 0.010, (iii) 0.990

Problem 2

(a)   0.0185

(b) $0.205T + 1/8760T + 0.001$

Optimum proof test interval = 8.4 days.

$fdt = 0.0106$

Problem 3

See fault tree opposite.

$P = 0.478$

Suggestions for improvements:

• duplicate cooling pumps;

• consider removing bypass;

• improve overall reliability of heat exchanger.

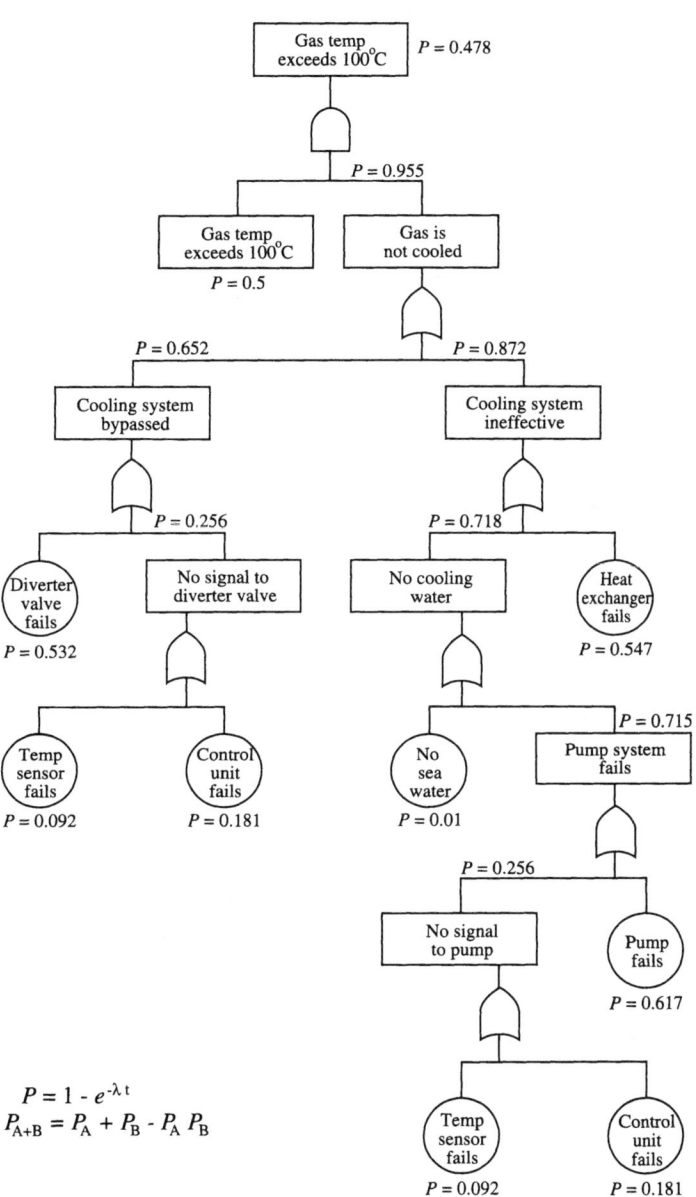

Gas temp exceeds 100°C $P = 0.478$

$P = 0.955$

Gas temp exceeds 100°C $P = 0.5$

Gas is not cooled

$P = 0.652$ $P = 0.872$

Cooling system bypassed

Cooling system ineffective

$P = 0.256$ $P = 0.718$

Diverter valve fails $P = 0.532$

No signal to diverter valve

No cooling water

Heat exchanger fails $P = 0.547$

Temp sensor fails $P = 0.092$

Control unit fails $P = 0.181$

No sea water $P = 0.01$

$P = 0.715$

Pump system fails

$P = 0.256$

No signal to pump

Pump fails $P = 0.617$

$P = 1 - e^{-\lambda t}$

$P_{A+B} = P_A + P_B - P_A P_B$

Temp sensor fails $P = 0.092$

Control unit fails $P = 0.181$

Fault tree — Problem 3

195

Problem 4

See fault tree below.

Approximate frequency of runaway reaction = 3.41×10^{-4}.

The cooling water system is the weak link; consider duplicate pumps and a cooling water supply low level alarm.

Problem 5

See fault tree opposite.

Frequency of boiling at HAL = 3.1×10^{-6} yr^{-1}.

Take steps to improve the reliability of the cooling tower and to ensure availability of fire pump in case of emergency.

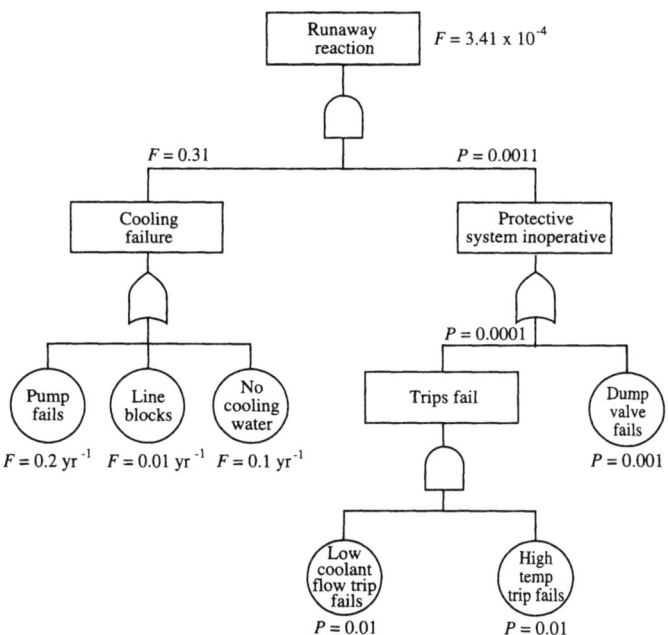

Fault tree — Problem 4

196

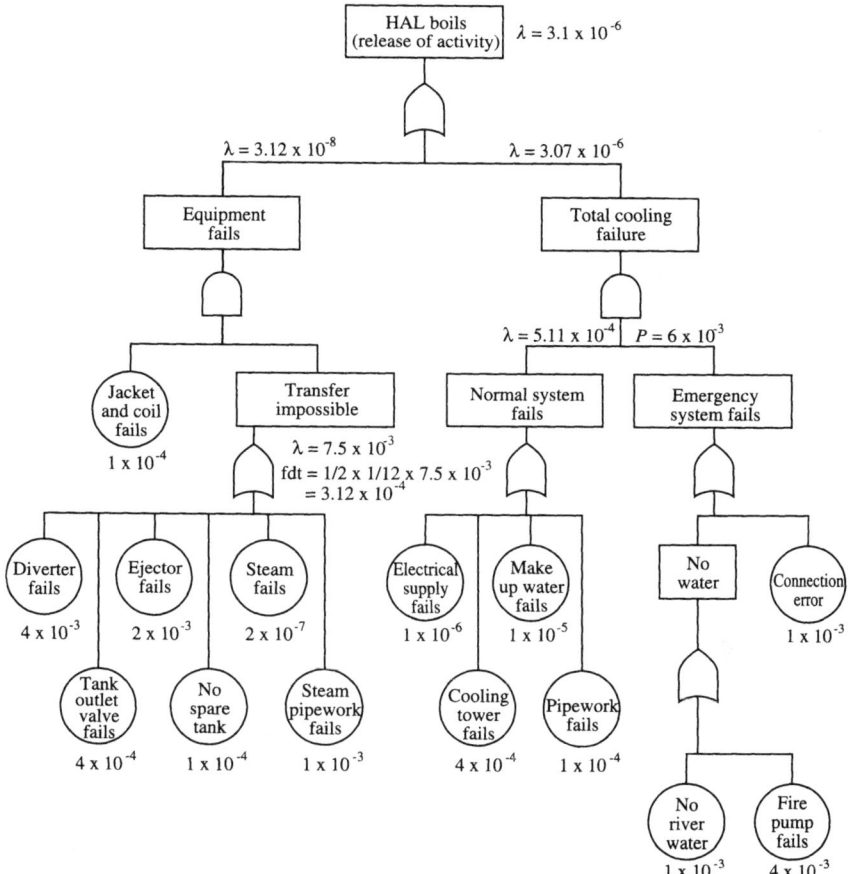

Fault tree — Problem 5

Problem 6
$\tau = 6.6$ days, $H = 6.82 \times 10^{-3}$ yr^{-1}, τ (two trips) $= 0.289$ yr^{-1}

Problem 7
Fractional dead time $\phi = 6.94 \times 10^{-4}$

Problem 8

See fault trees below. Hazard rates = (a) 0.05 yr^{-1}, (b) 0.043 yr^{-1}.

FAR is given by $0.043 \times \frac{1}{10} \times 50 \times 1000 = 215$ (totally unacceptable).

The LCV and positioner reliability must be improved and the process examined to reduce the frequency at which the reaction goes out of control. LT2 and LS2 should also be reviewed; are they really needed?

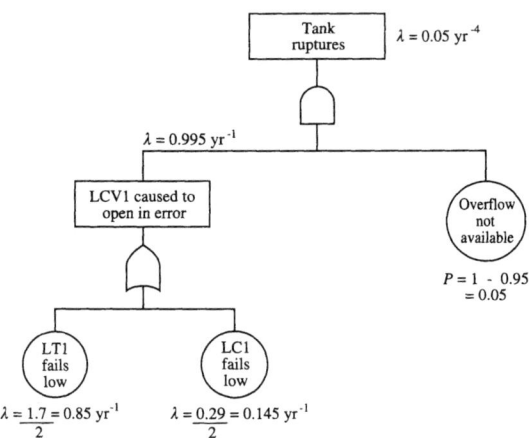

Fault tree — Problem 8(a). * Note that tank is initially full so LCV1 and its positioner are in the closed positions. Only an error in LT1 or LC1 will cause them to open.

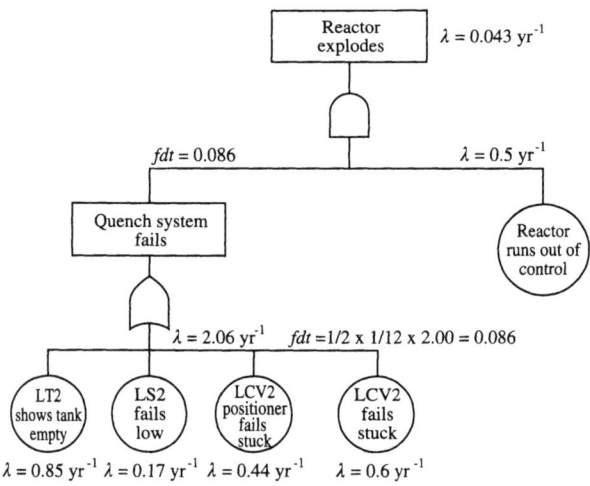

Fault tree — Problem 8(b)

Problem 9

See fault tree below.

Probability of top event $= 6.25 \times 10^{-6}$.

Adding a high level trip would give a combined *fdt* of:

$$\tfrac{4}{3} \times 0.0125^2 = 2.083 \times 10^4,$$

resulting in a top event probability of 1.04×10^{-7}.

Payback time (a) = 5853 years.

Payback time (b) = 11.6 years.

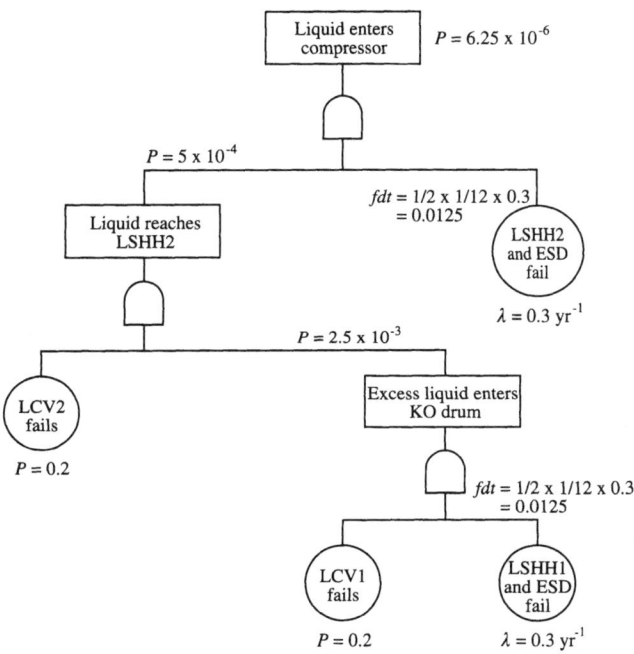

Fault tree — Problem 9

Problem 10

See fault tree below.

Frequency of explosion $= 2.3 \times 10^{-5}$ yr^{-1}.

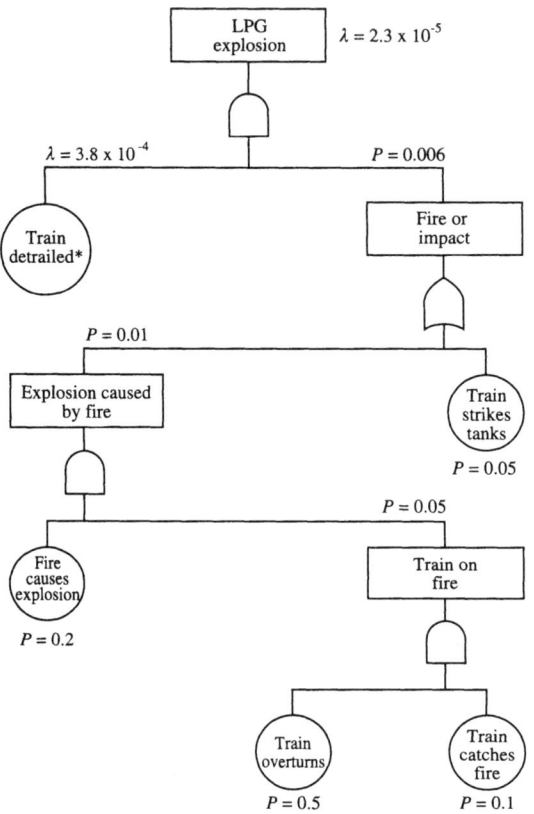

Assume that installation is at risk until half the train has passed clear of the last tank. Thus the total vulnerable track length is $0.7 + \frac{1}{2} \times 0.5$. Hence frequency of derailment is $0.95 \times 1000 \times 0.4 \times 10^{-6} = 3.8 \times 10^{-4}$ yr^{-1}.

Fault tree — Problem 10

Problem 11

See event tree below.

FAR = 0.4.

$$\text{Annual frequency} = \frac{0.4 \times 2000}{10^8}$$
$$= 8 \times 10^{-6}$$

Probability of death = 0.006.

$$\text{Maximum frequency of liquid breakthrough} = \frac{8 \times 10^{-6}}{0.006} = 1.33 \times 10^{-3}.$$

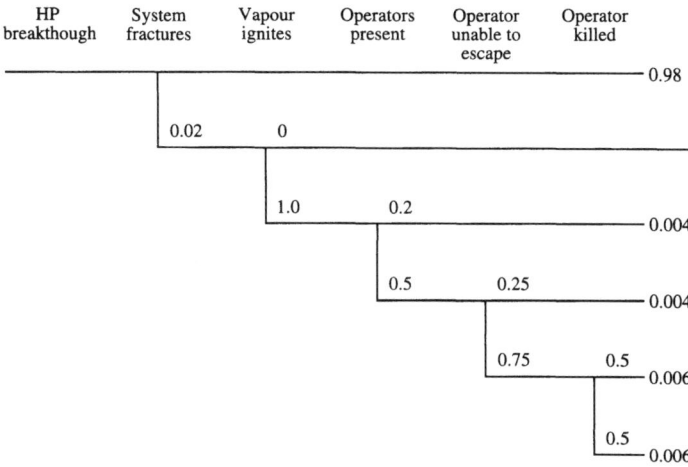

Event tree — Problem 11

Problem 12

See Tables 1 and 2. The FMEA in Table 1 shows that failure of the level control and trip system can have two effects; possible HP breakthrough and column high level. The latter is essentially an operational problem and thus has a lower consequence or effects rating than HP breakthrough. The analysis shows that it is possible that the LZAL and associated shut-off can fail and its failure would not be detected; hence the need for proof testing at regular intervals. Other failures

TABLE 1 (continued overleaf)

FMEA on distillation column bottoms system

Ref	Component	Failure mode	Local effect	
1	Level transmitter	Isolation valves shut	Spurious reading	
2		Impulse lines blocked	As above	
3		Probe stuck	As above	
4		Total	No output	
5	Level indicator	Total	No local reading	
6	LS1	Stuck high	No LAL	
7		Stuck low	Spurious alarm none	
8	LIC	Total	No indication or control	
9		No output	No control	
10		High set-point	High level	
11		Low set-point	Low level	
12	I/P converter	No output	CV1 shut	
13		Spurious output	No control	
14	SOV	Stuck shut	No response to LZAL?	
15		Fail open	CV1 shut	
16	CV1	Failed open	No control	
17		Stuck	No control	
18		Failed shut	No control	
19	CV1 bypass	Open	No control	
20	Air or power fail	Total	CV1 shut	

would still give a warning of possible HP breakthrough but operator intervention would be necessary. Because of the serious consequences of a HP breakthrough, a FMECA should be carried out — see Table 2, pages 204–205. This shows that the critical components are the bypass and CV1. It indicates that a full QRA should be carried out on the system. The results of the QRA actually showed that a second shut-off valve was required duplicating CV1 and removing the need for the SOV.

System effect	Indications	Comments/safeguards
HP breakthrough?	Low bottoms level	LZAL and shut-off
As above	As above	As above
As above	As above	As above
As above	As above	As above
None	No reading	Not significant
HP breakthrough?	None	LZAL and shut-off
	Continuous alarm	LI and LIC
HP breakthrough?	Low/high bottoms	LZAL and shut-off
As above	As above	As above
High level	High bottoms	Column flood
HP breakthrough?	Low bottoms	LZAL and shut-off
High level	High bottoms	Column flood
HP breakthrough?	High/low bottoms	LZAL and shut-off
HP breakthrough	None	Proof test needed
High level	High bottoms	Column flood
HP breakthrough?	Low bottoms	LZAL (no shut-off)
As above	High/low bottoms	As above
High level	High bottoms	Column flood
HP breakthrough?	Low bottoms	LZAL (no shut-off)
High level	High bottoms	Column flood

TABLE 1 (continued)
FMEA on distillation column bottoms system

Ref	Component	Failure mode	Local effect	
21	LS2	Isolation valves shut	No LZAL or shut-off	
22		Impulse lines blocked	As above	
23		Stuck high	As above	
24		Stuck low	Spurious alarm	
25	LZAL	No output	No LZAL or shut-off	
26	LG	Isolation valves shut	No indication	

TABLE 2
FMECA on distillation column bottoms system (Effect Category 1)

Ref	Component	Failure mode	System effect	
1	Level transmitter	Isolation valves shut	HP breakthrough?	
2		Impulse lines blocked	As above	
3		Probe stuck	As above	
4		Total	As above	
6	LS1	Stuck high	HP breakthrough?	
8	LIC	Total	HP breakthrough?	
9		No output	As above	
11		Low set-point	HP breakthrough?	
13	I/P converter	Spurious output	HP breakthrough?	
14	SOV	Stuck shut	HP breakthrough?	
16	CV1	Failed open	HP breakthrough?	
17		Stuck	As above	
19	CV1 bypass	Open	HP breakthrough?	
21	LS2	Isolation valves shut	HP breakthrough?	
22		Impulse lines blocked	As above	
23		Stuck high	As above	
25	LZAL	No output	HP breakthrough?	

System effect	Indications	Comments/safeguards
HP breakthrough?	None	Proof test needed
As above	As above	As above
As above	As above	As above
None	Continuous alarm	Check LIC or LG
HP breakthrough?	None	Proof test needed
No local level indication	No local indication	Obvious to operator

λ_p	x	λ_m	β	C_m
0.02	1.0	0.02	0.1	0.002
0.03	1.0	0.03	0.1	0.003
0.05	1.0	0.05	0.1	0.005
0.5	1.0	0.5	0.1	0.05
0.1	0.5	0.05	0.1	0.005
0.3	1.0	0.3	0.1	0.03
0.1	1.0	0.1	0.1	0.01
0.02	0.5	0.01	0.5	0.005
0.1	1.0	0.1	0.1	0.01
0.3	0.5	0.15	0.2	0.03
1.0	0.2	0.2	1.0	0.2
1.0	0.3	0.3	0.1	0.03
0.10	1.0	0.1	1.0	0.1
0.02	1.0	0.02	0.2	0.004
0.03	1.0	0.03	0.2	0.006
0.1	0.5	0.05	0.2	0.001
0.1	1.0	0.1	0.2	0.002

Problem 13
Minimum ventilation rate = 2084 m^3 hr^{-1}.
Operator has 5.73 hrs to LEL.

Problem 14
Overpressure at 50 m = 110 kPa.
Overpressure at 1 km = 1.5 kPa.
Casualty estimate at 50 m = 5.
Casualty estimate at 1 km = 0.

Problem 15
Emergency services can approach to 12.3 m.
Time to move injured person = 15 s.

LIST OF ACRONYMS

ALARP — as low as reasonably practicable

BLEVE — boiling liquid expanding vapour explosion

ELD — engineering line diagram
ETA — event tree analysis

FAR — fatal accident rate
FMEA — failure mode and effect analysis
FTA — fault tree analysis

Hazop — hazard and operability study
HSE — Health and Safety Executive (UK)

IChemE — Institution of Chemical Engineers (UK)

LEL — lower explosion limit
LNG — liquefied natural gas
LPG — liquefied petroleum gas

OEL — occupational exposure limit

P&ID — piping and instrumentation diagram
PFD — process flow diagram

QRA — quantitative risk assessment

UEL — upper explosion limit

INDEX